Confession Of A CIA Interrogator

Joseph B. Kelly as told to Ben R. Games PhD

www.FideliPublishing.com

This book is a work of non-fiction. Unless otherwise noted, the author and the publisher make no explicit guarantees as to the accuracy of the information contained in this book and in some cases, names of people and places have been altered to protect their privacy.

©2009 Joseph B. Kelly as told to Ben R. Games PhD. All rights reserved.

No part of this book may be reproduced, stored in a retrieval system, or transmitted by any means without the written permission of the author.

First published by AuthorHouse 1/3/2007

ISBN: 978-1-60414-197-9

Library of Congress Control Number: 2006910512

Printed in the United States of America
Bloomington, Indiana

This book is printed on acid-free paper.

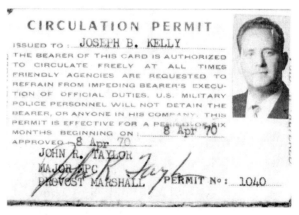

Vietnam CIA Circulation Permit

DEDICATED

TO

THE CENTERAL INTELLIGENCE AGENCY (CIA)

COMPANY PATRIOTS

by

Montana, GD, Security Chief,
The Games Clan

"Confession of a CIA Interrogator"

Author's Statement
About the Cover Picture
Facts
Prologue
Moriggia's Confession
Historical Documents

www.GamesLittleBigBOOKS.com

As told to Ben R. Games, PhD by Joseph B. Kelly, Vietnam Police Medal of Honor

---

Vietnam pictures by: Helen M. Games, MBA, Joe B. Kelly, and Dave L. Walker. Edited by Sarah H. Kelly

STATEMENT

This is a semi-biographical story of a CIA Interrogator. The facts of the story are based upon information from "Tour of Duty" of Senator John Kerry, (D) in Vietnam by Douglas Brinkley, Hanoi Jane's Apology in "The Washington Times", the biographical account of Senator John McCain, (R) as a prisoner in the "Hanoi Hilton", talks given by other prisoners, Sergeant Dave L. Walker (25th Infantry Division) diary, the author's Vietnamese Journal entrees, interviews with Lt/Col Geoffrey T. Barker, US Army (Ret) and Joseph B. Kelly's adventures as Gilbert H. Moriggia in Vietnam.

History is a window that looks into the future. Today we live in a Geo-Feudal Period. Instead of Kings, there is CEO, Instead of Knights there are Managers, and instead of Serfs, there are Stockholders who are loyal to geo-companies without national boundaries.

The reader should study the Bible's book of Genesis and Confession's Historical Documents to verify for themselves what part of this story is a historical fact. There is no plot or conclusion to this adventure. The tale is only about the period when the Company Interrogator Gilbert H. Moriggia existed during the Vietnam War.

Ben R. Games, PhD

CIA Contract Agent (Gilbert) & PRU Tiger Scouts

CAPTURE OF A PRP MEMBER

In Gilbert's office there hung a large map of the Kien Phong Province. It was on the wall across from the office's only door showing all the Hamlets, Villages, and Cities. Weekly and daily reports from police informants were helping to track the communist cell member's movements, which were then marked by pins, stuck into the map. The three members were leaders of the Peoples Revolutionary Party (PRP) and were represented by three different colored pins. They were continually on the move; never traveling together.

Some historians claim that no one was able to capture a Communist PRP member. Like most of the terrorist news releases to the wire services and media, lies are mixed in with a little truth. Everyday Gilbert would study the map trying to find a pattern to their movements. Then one day as he sat studying the map he jumped up, went over, and started drawing a line tracing the movements of the red pin. There was a pattern as the pin was always moving along roads or trails away from the Mekong River. Yesterday the pin had stopped in a village where some of the children had disappeared with the Viet Cong guards. The people were scared but the children were not in school and the informant, who was one of the teachers, reported them missing to the police. The village was a night's walk from three other villages with only one near the river. From the informant Gilbert now knew that the red pin represented the PRP male member. If the man had walked all night, he would still be three hours from any village. Now all Gilbert had to do was guess which of the three villages he would be visiting.

All the police intelligence officers were giving their reasons for one of the two villages located on land away from the Mekong when Gilbert spoke up saying, "He's headed for the village nearest the river". No one agreed but Gilbert looked at his watch, and told the radio operator to contact the S-2, Chief of the Tactical Operations Center (TOC), and have a Slick (UH-1) pickup his Province Recon Unit (PRU) Tiger Squad. They had forty-five minutes to get in position for an ambush to make the snatch and grab.

Ten minutes later a US Army Slick (UH-1) made the pickup. Gilbert, a Navy Seal Advisor, and the PRU Tiger Squad were on their way to try for the first arrest of a member of the Peoples Revolutionary Party (PRP) in Vietnam.

Gilbert directed the pilot to fly over the Mekong River away from the city, and then had the helicopter set down near a dirt trail that ran past the ambush site a ¼ -mile from the town. He asked the pilot for a fast drop-off, then to fly outbound away from the river for five miles, and wait for a radio call to return. The Navy Seal suggested that the PRU Tiger Squad hide in the reeds close to the riverbank near the trail. Not a good tactical position but Gilbert knew that the snatch had to be quick and fast with absolutely no warning or the PRP member would have time to commit suicide.

He had guessed right. The man came walking down the trail talking to his two bodyguards, carrying AK 47 weapons, who were watching the trees lining the trail. There were twelve PRU Tiger Scouts in the squad with Gilbert led by a Navy Seal hiding in the reeds. As he came abreast of Gilbert, the PRU Tigers Scouts rose out of the reeds and grabbed the man before he knew they were there. As they pulled the PRP member off his feet, five of them were pulling at his clothes while grabbing for his weapon. They tumbled about rolling toward the river with a waving mass of arms and legs, stripping him of everything that he could use to commit suicide. Suddenly gunfire erupted and just as quickly stopped. The surprise had been complete. Only the Viet Cong guards had been killed. Gilbert was on the radio calling for pick up, and someone popped red smoke. The Slick arrived, and they were gone taking the enemies weapons with them. As Gilbert looked down at the reeds from the helicopter, they were already standing tall with no sign that anyone had been there except for the two dead bodies lying in the road.

The cover picture was taken by Gilbert upon landing at the National Police's home base in Cao Lanh. Notice the smiles on the faces of the PRU Tiger Squad for they had just made the first arrest of a Peoples Revolutionary Party Member. The prisoner in the white shirt is a Communist PRP Member who was immediately flown to the National Interrogation Center in Saigon.

All Province Recon Units (PRU) Tiger Scouts were men recruited from the Viet Cong (VC). When they operated at night they wore black uniforms nicknamed "Pajamas" by the Americans, and during the day, it was striped camouflaged uniforms giving them the nickname of Tigers. There was no written rule of which uniform to wear or when, and some wore the black pajamas day and night. Gilbert thought it was the black hats they liked.

FORWARD

The writer has been diagnosed by the Veterans Administration as having Myasthenia Gravis (MG) which is affecting his eyes. He is now a one finger typist and can not read paragraphs. Montana, GD, is a male black Labrador Retriever Southeastern Guide Dog, and companion to Helen Games, MBA. He has helped the writer Ben Games, PhD, investigates and research stories for the past nine years, and believes all the mistakes in this story are computer errors.

The story is Non-Fiction and covers a period of 2½ years of Joseph B. Kelly's work in Vietnam. Joe became a CIA contract agent assigned to assist the US Army as a civil servant GS-11 police advisor for South Vietnam. The writer requested Joe to provide a few pages of the story to Mr. Ken Tolliver, CIA Retired Jeweler, and Company man to get his comments.

This is what he wrote; "Some called themselves mercenaries, others termed them heroes, but to most they were dangerous men who had entered a nether world of shadows. They had surrendered their identities. T.S. Eliot would call them "The Hollow Men". Officialdom had a more bureaucratic title; contract agents.

Contract Agent was a sanitary and convenient catch-all description. Their lives and usefulness were neatly, if often vaguely, spelled out in written descriptive documents hidden away in secret archives. The government agencies could pay them almost any wages; they were often provided military titles of convenience which allowed the accounting offices justification for their remunerations. Further these agreements, or contracts, defined the elaborate controls which assured discipline and obedience. These controls were cunningly constructed and utilized both punishments and rewards. A non-commissioned officer might be brevetted a Colonel to give him authority, and they could be awarded medals to provide prestige or feed their ego. But, if they violated their contractual obligations, they forfeited pensions; they got no references, and became non-persons. At the end of their contracts most found that they had lost families and ties of past friendships. All of their personal history was hidden and they were forced to seek empty livelihoods. Some drifted into the service of foreign flags, most discovered that earning a living without a respectable past was impossible.

The great majority became what psychologists called "frozen in time." That is, the few months or years they fulfilled their contracts locked all their future days in a secret vault they could not reopen. Often, even recognition of the names under which they lived and operated as part of their identities was denied them.

Air Force Technical Sergeant Joe Kelly retired honorably in 1968 after 20 years of military service. He should have taken his wife and perhaps attended VFW meeting and savored memories of his family life back in the United States. But in far off Vietnam, as the result of a tragic change in US foreign policy and the fateful decision to back the return of the French as a colonial power in Southeast Asia, the United States was drawn into a bloody and protracted war, condemned even by the French. This new conflict, fought largely by irregulars and secret armies, demanded fighters with special skills.

They would serve operations controlled by a new set of initials, CIA, as contract officers. As with most covert operatives, his own identity was the first casualty of the "contract". Joe Kelly became Gilbert H. Moriggia, a special police advisor in Kien Hoa Province, South Vietnam. His civilian title veiled his more meaningful role as CIA Interrogator and Intelligence Officer. In fact, his pseudonym cloaked his vital mission. His job was to plan and execute the kidnapping of high ranking enemy civil and military personages. Such people could prove a wealth of intelligence and if "turned" could become prized assets.

The nation's top spy organization selected the retiring Air Force non-com for this operation because Joe Kelly; offered unique qualifications, coupled with equally specialized training that would transform him into Gilbert, the contact agent. His qualifications included **(OK Joe, insert your resume here. Make it unique, and justify your hiring.)**

His training period consisted of (okay here describe in detail your progress through Blue U, etc. Be careful about any secrecy violations, but take some risks…after all, the readers want to know how "spies" are trained. **(This chapter dealing with your recruitment, training, and preparations should run a 100 pages!!!)."**

FACTS

Gilbert H. Moriggia, CIA Interrogator, Intelligence Officer, Special Police Advisor Kien Hoa Province. Created by the CIA on 5 August 1968 and removed from service on 2 December 1971. Gilbert is a CIA Pseudonym or alias for Joseph B. Kelly, T/Sgt USAF. After 20 years 6 months military service he was retired from the USAF, and became Gilbert. That's a FACT.

The Province Recon Units (PRU) with their Navy Seals or Special Forces Advisors captured Viet Cong (VC) for Gilbert to interrogate. The PRU Tiger Scouts were transported on recons by Navy Swift Boats (PCF) including one captained by a Navy LTjg, and by US Army Helicopters with one commanded by CW-3 Ben R. Games (Gentle Ben). That's a FACT.

South Vietnamese PRU Tiger Scouts were Viet Cong who were recruited for Province Recon Units. They were paid by the CIA to inform on other VC and to capture prisoners for interrogation. Gilbert as a member of Advisor Team #88 was wounded twice in firefights by the Viet Cong (VC) while working with the PRU Tiger Scouts and Special Police Forces. He was treated at the US Army MACV medical facility, and the second time in the emergency room at the Tuy Hoa Air Force Base Hospital. That's a FACT.

During an interrogation of prisoners, Gilbert identified the Vietcong Cell Leader of the Zapper Squad (Terrorist) in Ben Tre City. She was Nighi Thi Nruyent the daughter of one of the North Vietnamese General Officers at the Paris Peace Talks. She was transported BLACK to Paris and allowed to phone her father just before the meeting that day. That's a FACT.

Joseph B. Kelly was awarded the Police Medal of Honor by President Thieu of South Vietnam with a military ceremony in Kien Hoa Province, 22nd June 1970. He also received the Cross-of-Gallantry w/Bronze Star, the first Pacification Medal of Honor, and an award from General Creighton Abrams, US Army, for assisting the United States Army. That's a FACT.

Gilbert (CIA Pseudonym for Joesph B. Kelly) spoke Vietnamese, Japanese, Spanish, Italian, French, and German. God blessed him with a talent for remembering small details about people and their habits. That's a FACT.

Little Big BOOKS have large print so everyone who needs a little help reading can enjoy the adventures of Gilbert the CIA Interrogator. This story is actually a biography of Joseph B. Kelly's service for his country during the Vietnam War. This was a war where US citizen soldiers never lost a fight but the people at home gave up, and the war was lost. That's a FACT.

Ben R. Games, CW-3, Gentle Ben, Aircraft Commander of a CH-47 Chinook helicopter flew for Co B, 228th Aviation BN and E Battery, 82nd Field Artillery, 1st Cavalry Division. Awarded the Distinguished Flying Cross for heroism, Bronze Star, and 13 Air Medals while flying combat in Vietnam. He was wounded in a firefight at LZ Vivian. Ben's special missions were to fly Vietnamese Province Recon Units (PRU Tiger Scouts) to capture prisoners for interrogation and Artillery Raids into Cambodia. He and his wife Helen lived behind the American PX in an Australian 26 foot Willerby Caravan on a firebase called Bear Cat, Vietnam. That's a FACT.

The US Central Intelligence Agency often referred to as the Company, or just as the CIA, operates all over the world gathering intelligence about terrorists and their operations in an effort to protect the American people. Terrorists and Dictators have a deep fear of the Company and of the patriots who work within its organization to protect America. During the Vietnamese War enemies of the United States knowing that they could not win a fight with American soldiers on the battle field used terrorism, suicide bombers, and electronic communication to flood the media channels with an avalanche of lies to destroy America's will to defend itself. That's a FACT.

Most of the prisoners in the Hon Lo Prison-- infamous "Hanoi Hilton" were US Navy pilots and aircrews who had been shot down while bombing targets in North Vietnam. By the end of the war, there were over 500 American prisoners. The interrogations started within minutes of their capture. It did not matter if they could not understand Vietnamese. The prisoners would have their hands wired behind their back with the elbows tied together. Sometimes a rope would be tied to their wrists, and while two soldiers held the American pilot another one would pull on the rope until the shoulders were pulled out of their sockets. As a prisoner Navy pilot, John McCain was hung up by his two broken arms and his teeth knocked out. Then he was beaten day and night for a week. This was the start of Interrogation North Vietnamese style. That's a FACT.

One North Vietnamese prison guard (Americans called them "V") was an interrogator named "Rodent" who was trying to obtain military information. He would put a rope under the armpits of the prisoner, stand him on a chair and then pull it away. The rope stopped the prisoner with a jolt. After swinging in the air for two hours, the questioning would begin. First, they would twist one arm until it broke and then ask a question. If the prisoner was captured two or three days from Hanoi, he would be forced to walk or be dragged with stops along the way to allow villagers to see and beat the American. The "V" torturers had three basic kinds of torture--kneeling, ropes, and beatings--each of which had numerous variations. That's a FACT.

When Jane Fonda visited North Vietnam to interview American prisoners all the prisoners would know weeks before her arrival. Frenchy, Commandant of Hanoi Hilton would prepare a script for each interview and ask for volunteers. The Americans would have their hands wired behind the back and forced to stand in the middle of their cells barefooted with no food or water while they waited for Hanoi Jane. If they moved or fell, they would be beaten with clubs. This would continue until some one volunteered to let her interview them. That's a FACT.

Gilbert was interrogating Viet Cong prisoners to obtain military intelligence in an effort to save the lives of American fighting men. Jane Fonda was also busy planning ways to get the Communist government to allow her to interview American prisoners. That's a FACT.

In a speech to 2000 students at the University of Michigan on 21 November 1970, she told them, "If you understood what Communism was, you would hope, you would pray on your knees that we would someday become Communists." At Duke University Jane Fonda repeated what she had said in Michigan, adding, "I, a Socialist think we should strive toward a socialist society, all the way to Communism." That's a FACT.

With constant interrogations occurring at any time day or night for as long as seven years some men stopped eating just to die. When the other prisoners were near them as the rice and chicken heads were issued they would hold their buddy down and force food in his mouth. If he threw up, they would gather the vomit and force it in his mouth until he swallowed. All those who were treated to this forced feeding lived to return home. The North Vietnamese did not return some prisoners, and these became MIA's. Many

MIA warriors were prisoners who had been crippled by torture beyond belief and some had become insane. That's a FACT.

After WW-II French Indo China was returned to France by the United Nations. While attending the USAF Air University Nuclear Weapons course, Gentle Ben attended a secret briefing where the Intelligence Officer explained that the French would leave Vietnam. Then he said that anyone using conventional weapons that killed 10,000 of the enemy per month for ten years would lose the war. By North Vietnamese count, twenty-five percent of their population was lost during the war. That's a FACT.

This was a Communist lie for when a soldier left North Vietnam heading south carrying a rocket to fire into a city of South Vietnam he or she was discharged from their Army. After walking to their target area and firing the rocket, they would join the VC or sometimes a South Vietnamese Province Recon Unit (PRU). All those joining the PRU were paid approximately $200 dollars per month. There were no bonuses paid for killing anyone. The PRU Tiger Scouts were used to capture prisoners for interrogation attempting to prevent future terrorist attacks. The only thing the VC and PRU had in common was that they never returned to North Vietnam. The enemy claimed that they never invaded South Vietnam, and that America was the aggressor. Using this method of accounting they were able to rid North Vietnam of all those who believed in Jesus Christ. That's a FACT.

As he flew over Vietnam Gentle Ben, would climb his CH-47 Chinook helicopter to over 2000 feet and circle around some towns to stay above small arms fire. When landing at these towns while picking up PRU Tiger Scouts he would have his gunners' recon by fire. When questioned by some of the 1st Cavalry Company Commanders on how he knew which towns to avoid he would answer telling them he could tell just by looking at the town's churches, and temples. That's a FACT.

After the Vietnam Peace Accord (Paris Peace Talks) was signed 27 January 1973 (with cease-fire on 28th Jan. 1973), the Americans left the Vietnamese people to fend for them selves. Many American newspapers reported that it was an unjust war and not our problem. When the North Vietnamese took command of the South Vietnamese soldiers there was no food issued for the next seven days to any soldier of the Christian Faith. Then on the 7th day, all the parents and relatives were asked to come to the military parade ground to get their sons. The men were lined up at attention while the families watched. Then came an order to space the ranks three paces apart and dress right. A North Vietnamese Officer marched down behind the ranks with his pistol. He would stop at each Christian Soldier raise his pistol, and shoot the young man in the back of the head. About every third or fifth solder was killed in this manner. That's a FACT.

Many Americans still do not believe it was wrong to give up or understand the hate the North Vietnamese had was not that the pilots captured were US Navy but that they were Christian warriors. The South Vietnamese families still do not hate Americans for leaving and letting their sons be killed but they cannot forget that the young men were starved first. Guests of the Hanoi Hilton may forgive but will never forget Jane Fonda, or the six prisoners who aided the enemy. That's a FACT.

Gilbert H. Moriggia learned fast from the Viet Cong (VC). He never went to his office in Ben Tre City over the same route or at the same time. He learned never to eat in the same restaurant twice on the same day, and never become predictable. Always sit with a wall at your back and do not drink the water or use ice. Most restaurant owners wanted him to sit where he could not be seen from the street. The Viet Cong (VC) liked to drive a Honda 50 past a restaurant, and if they saw an American would throw in a grenade for an appetizer. That's a FACT.

Joseph B. Kelly, CIA Case Officer, reported to the CIA Province Officer In Charge (POIC) of Kien Pong Province, Caesar J. Civitella, and later worked directly for William (Bill) Colby, Directorate of Plans, Policy, & Programs (MACORDS) on an operational plan for interrogations that developed into the first Dai Phong Program (Hurricane or Big Wind) used in Vietnam. In the first four months of its operation, the program eliminated more Viet Cong (VC) than all other military operations combined in Ben Tre City, Kien Hoa Province. That's a FACT.

Ben Tre City, Kien Hoa Province, the birthplace of the Communist Viet Cong (VC) was the area where Gilbert met the terrorists head on. After four months of constant pressure by Gilbert, the Vietnamese Department of State Province rated the city as pacified. That's a FACT.

Joseph B. Kelly, born 1930 Darlington, MD, has eight brothers and sisters. He joined the USAF 2 Jan 1948, served 20 years five months and retired 1 May 1968. Was a parachute rigger and worked for special ops USAF high altitude aircraft. Recruited as a CIA Case Officer GS-11 on 1 Jan 1968 and assigned to Vietnam in charge of PRU-SCG-RDC. He was wounded twice (6/8/70 Ben Tre City & 9/22/70 Tuy Hoa city). The firefights were classified secret and no Purple Hearts were awarded. The Vietnamese government awarded him the Cross-of Gallantry w/Bronze Star, Pacification Medal of Honor, Police Medal of Honor presented by President Thieu, and the Civil Service Medal by USA General Creighton Abrams. That's a FACT.

Joe worked on the Dai Phong Program with Captain Geoffrey T. Barker in Ben Tre City, and was then assigned by Bill Colby to implement the program in all the areas along Hwy 7 and 14 to Cambodia wherever the Viet Cong was attacking the South Vietnamese people. He helped defeat the Communists in Tuy Hoa, Nha Trang, Binh Dinh, and Pleiku. That's a FACT.

CW-3 Ben R. Games (Gentle Ben), 1st Cavalry Division, wounded in a firefight at LZ Vivian returned to Fort Rucker, AL, for medical treatment on the 5th February 1970. The First Cavalry Division battle for Hwy 1, on 31 May 1967, seized Luoi airstrip renamed LZ Stallion 24 April 1969, Battle of LZ Carolyn 6 May 1969, Firefight LZ Vivian 8 November 1969, entered Cambodia 1st May 1970, stood down from combat on the 28th March 1971, and all military forces left Vietnam on 28 March 1973. That's a FACT.

Some of the events may not be in the correct chronological order but were recorded as research discovered them. That's a FACT.

PROLOGUE

The Garden of Eden must have had a temperature averaging 80 degrees throughout the year with an annual rainfall of 79 inches. Fuujin's summer winds would bring the rains down from the north cleaning and sweetening the air. Tropical rain forests surround the garden leaving the Mekong delta and plains the richest rice-growing area in the world to feed God's children.

This is South Vietnam, a land torn between God and the Devil. Into this land came three people unlikely to ever meet, but destined to learn from each other. One man was from the US Navy a LTjg Skipper of a Swift Boat, one was an Army Chief Warrant Officer, Chinook Air Craft Commander, and one was a USAF T/Sgt retired who became Gilbert the CIA Interrogator.

During WW-II, this peaceful land was known as French Indo China, and was invaded by the Imperial Japanese Army. A few years after Japan was defeated the United Nations divided the Nation into three countries called Laos, Cambodia, and Vietnam. France was given control of Vietnam until the Communists forced them to retreat to the southern part of the country. All Vietnamese Christians were then forced to denounce God and become Communists or leave their homes for the move south where a new country named South Vietnam was created in 1954 by the United Nations who established a Demarcation line at the 17-degree Latitude.

The South Vietnam capital of Saigon was a beautiful city. It soon became known throughout the world as the Paris of the Orient. Only ten percent (10%) of the population was of the Christian faith with all civil service employees, ministers, and leaders being Catholic. The population of South Vietnam in 1968 during the time Gilbert the CIA Interrogator worked in Vietnam was 15,715,000 people living on 65,726 square miles of land (similar in size to the state of Washington).

The land was controlled by two forces. The Dark Force known as the Viet Cong (VC) under the command of the Communist "Peoples Revolutionary Party Central Committee (PRP)" of one man and two women supported by the North Vietnamese, and the Bright Force under the command of their elected President Thieu supported by the Americans. The Dark Forces controlled the country after the sun went down. They set Booby Traps on trails, roads, and bombed government facilities. They also practiced forced recruiting of young people for their fighting cadre. Boys and girls of ten years old made good soldiers and had no fear. There were Communist tax collectors with police and courts. The nights belonged to the terrorists.

During daylight, the Bright Forces took control. Children could play without fear. The Army helicopter gunners would recon by fire when approaching a landing zone (LZ) in enemy territory. Army gunners would not fire at children but would fire at the bushes or buildings on the LZ. Whenever a child was accidentally wounded, they would have a PRU Tiger Scout or Vietnamese soldier search them before they boarded a medical helicopter. Sometimes a child would be carrying a live grenade with the pin pulled.

The Viet Cong had another device they used against the Navy. It was a large woven basket made waterproof with tar that could carry a bomb under it. A child would be placed in the basket with a trip wire that would explode the bomb against a ship or boat when they were lifted up. The Navy had to sink the baskets with cannon fire before they were close enough to see inside.

When Gilbert the Interrogator was sent to the Mekong Delta region, he was given a mission to assist the Province Recon Unit (PRU) Tiger Scouts and Special National Police in establishing an intelligence system to find and locate terrorists. To accomplish this mission all the Dark Force Tax Collectors and other officials of the Communist Viet Cong had to be identified. He desperately needed to capture and identify the enemy, not kill them. Gilbert could not interrogate or find out where they were going to attack if they were dead, and nothing was gained if they were allowed to commit suicide. Some of the 720 young men in Gilbert's Dai Phong Program were PRU Tiger Scouts attempting to stop the VC terrorist attacks. As members of the PRU Tiger Scouts, they were paid a monthly salary of about two hundred dollars, and the Communist Viet Cong women as paid informers received $25 for weapons and $50 to identify VC officials.

To capture Viet Cong terrorists for interrogation Gilbert recruited other VC into joining the PRU Tiger Scouts. To assist him the Tactical Operations Control (TOC) assigned Navy Swift Boats and Army helicopters and trucks for his attempts to capture prisoners for interrogation. Into this mix came a Navy LTjg skipper, and Army Aviator CWO Gentle Ben.

MORIGGA'S CONFESSION

CHAPTER ONE

On its border with Laos, the South Vietnam high country is dominated by the rugged Annamese Cordillera mountain chain with fertile deltas along the coast. The central area consists of a vast plateau while the south is dominated by the Mekong delta and plains. Into this arena Gilbert, the CIA Interrogator, was busy recruiting a Viet Cong (VC) force of PRU Tiger Scouts to capture Communist Viet Cong for Interrogation.

Along the many mouths of the Mekong River in the China Sea up to the Cambodia border Navy Swift Boats carried Gilbert's PRU Tiger Scouts in their raids trying to capture Viet Cong for interrogation. It was always to capture the enemy and not to kill for Gilbert had no need for dead informers. It was only the live prisoners who could help save the lives of American soldiers.

During the day, farmwomen working in the rice field wore large straw hats, and dark slacks with a dark blouse or shirt. Their men wore a similar outfit and from a distance looked the same. In town, the men wore dark pants and white shirts. With their lean narrow waists and straight backs, they could be called handsome. The women wore loose white silk slacks with split blouses that went down to their ankles in orange or other bright colors and looked like beautiful fluttering butterflies when they walked.

Unlike the Monsoon rains that clean and refresh, these dark clouds were Communist Viet Cong coming down the Ho Chi Minh Trail bringing death and destruction to this peaceful land. Vietnam Shangri-la had a serpent within its borders who professed love and peace in the daytime with an appetite for terror against the people when darkness fell over the land.

Aboard a TWA plane on its final approach to the Tan Son Nhut Airport, Joe Kelly sat and practiced saying to himself, "Gilbert H. Moriggia is my name." He could see the city of Saigon out of the planes' windows as the pilot banked the DC-6 lining up for the landing. As the airliner shot past the approach end of the runway, he caught a glimpse of fighter aircraft, carrying bombs under their wings, waiting to take-off. As they taxied into the terminal, he could see that a rain shower had just passed, and puddles of water were still standing on the walkways. The sun was shining, the air was clear, and everyone was busy. Trucks, freight, and people were going every which way. Gilbert thought to himself, "I wonder where the war is?"

Five days after retiring with over 20 years as a USAF T/Sgt Joe Kelly signed a contract to become a CIA Contract Agent and along with thirty-five (35) other men reported to the Central Intelligence Agency Training School in Virginia. Besides attending language classes in Vietnamese, they were taught how to become Intelligence Officers (GS-11) working for the US Government. Two months latter he became Gilbert assigned as the CIA Agent in charge of the Revolutionary Development Cadre (RDC), (the Vietnam Peace Corps), the Station Census Grievance (SCG), (the paid informants for intelligence information), and the Province Recon Units (Viet Cong working for the CIA) in Cao Lanh, Kien Phong Province.

The trip to Saigon was on a scheduled TWA Airliner flight from the United States with stops in Honolulu, and Hong Kong. The passengers were traveling on business and pleasure. Before the plane started its landing approach to the Tan Son Nhut Airport near Saigon, one of the hostesses went down the rows of seats asking for passports from the women and some of the male passengers who were getting off in Saigon. CIA Intelligence Officers and Diplomatic Couriers were not questioned or asked for passports. He had noticed that most of the men leaving the plane were armed, and guessed that most of the women were just as deadly. A Vietnamese Immigration Officer was first to board the plane after the landing, and while standing in the doorway went through the passports very quickly. The hostess returned the men's passports as they disembarked, and told the women that their passports would be returned inside the terminal. Gilbert did not wait around but joined two men he had met in Virginia who were waiting at a sign that read BUS. It was a special bus that took them directly to the Hotel Duc, which turned out to be a hotel for Company Agency (CIA) people.

Gilbert first view of Saigon was in a traffic jam as bad as any he had seen in Paris, France. The streets were crowded with bicycles and motor bikes. They were twelve abreast; handle bars to handle bars going one way with the same number traveling in the opposite direction. There were also motor scooters, Honda 50's, Army jeeps, trucks, and their bus in this mix of traffic.

The streets were wide boulevards lined with trees boarded by shops and buildings of not more than five floors high. There were no skyscrapers or tall buildings, and only the Catholic Church's steeples rose above the city.

In the market area and side streets the shops were owned by Chinese, Indian, and Koreans who dominated the area. Among these shops and on the sidewalks were Vietnamese merchant's stalls selling everything anyone would every need.

While the driver inched the bus along Gilbert watched a young American Red Cross worker trying to start her motor scooter. She was standing in the street pushing the starter crank with her foot when the engine started. She had hold of the handlebars but was not on the scooter. It went around and around with her hanging on until it hit one of the bikes. The rider fell over hitting the bike next to it, and then fell into the next one. Like the slow motion, falling of dominos all twelve of the riders fell over into the street, and behold somehow the next row traveling in the other direction started falling to the pavement. There was no sound in the bus as the rows of bicycles continued to fall in slow motion until someone started laughing.

The bus stopped, and before Gilbert laid a street full of people lying on the pavement with their bicycles. The lady in the Red Cross Uniform pulled her motor scooter over the curb onto the sidewalk and drove off. Everyone on the bus including the driver cheered as she passed them. The riders got up, brushed each other clothes, and were laughing as they got back on their bicycles to wait for another green light.

The plane had landed at ten that morning, and after three hours, the bus driver told him that the hotel was only a twenty-minute drive from the airport but that they would soon be there. Gilbert was watching, studying, and trying to learn about the Vietnamese people. This land was different but the people could laugh at crazy mistakes just like at home. He thought to himself that they must be worth helping. The bus still did not arrive at the hotel until the traffic jam was cleared and that took another half hour.

As he walked into the lobby, he noticed men sitting in the foyer with bandaged injures; one had both arms and one leg in a cast. It looked as if the bus driver had made a mistake and dropped him off at a hospital. The front desk was to the left of the entrance and part of the lobby. A grand wide stairway lead up to the first floor of rooms directly across from the entrance, and anyone going up to the rooms or floors above had to pass in front of the desk. It was a small hotel exactly like those in many small towns of France. It was easy for Gilbert to imagine that he was visiting Metz or Nancy, France except for the six-foot high steel fence surrounding the hotel.

The Duc Hotel was near General Creighton Abrams quarters and within walking distance of the American Embassy. It was also within a few blocks of a Vietnamese government office building, which was a block long and two stories high. The street was blocked with large concrete barriers, and everyone entering the area was searched. Vehicles were parked in a guarded lot where the guards walked around pushing mirrors mounted on wheels looking under the vehicles for bombs.

Terrorism was a way of life, and when a VC was caught or observed planting a-bomb or booby trap, they were shot on the spot. There was no such thing as giving a terrorist a second chance or even letting them entertain the Americans by preaching about Dark Force pleasures.

There were phones in the Company's hotel but you could not call to the United States. The desk clerk told him that the Eden Roc Hotel on Tu Do Street had phones where you could call home. He got a ride to the hotel one afternoon while waiting for his CIA in country briefing. The Eden Roc was a VIP hotel that catered to news reporters, General Officers, and Very Important People (VIP) from all over the world. On the second floor was a room where anyone could place a phone call to the United States. When Gilbert walked into the room, he noticed six wooden phone booths lined along one wall with a woman in a Red Cross uniform setting at a table. On the table was a sign reading $12 US per minute. Gilbert walked over and signed the register as Joe Kelly paying for three minutes. He noticed there were only six callers ahead of him. An hour later his name was called, and he was told that his party was on the phone in booth number three. After the phone call was finished, he decided to stay in the hotel instead of returning, for tomorrow he would be starting his CIA briefings on the third floor of the American Embassy.

The restaurant and bar of the hotel were on the top floor with rooms on the four floors above the lobby. The roof was a floor of colored tile that had table and chairs where guests could visit friends. At night there were no lights except for the flickering of candles, and the occasional flash of light coming through the stairway door leading to the bar area. After visiting the stores along the street and market area Gilbert returned to the hotel. He went up to the lounge and soon discovered that the hotel was also used by Company men on R&R, news people, business men, military VIP's, with more American, and French women than he'd seen in one place outside of Paris, France.

Later, after night had fallen he walked out of the hotel onto the sidewalk. It was hard to walk with people sleeping in cardboard boxes against the curb. The weather was warm so it wasn't the cold people were afraid of but the sound of incoming mortar rounds hitting nearby. Once or twice gunfire erupted a block away but no one moved or left their box. As he walked along studying the sleeping families he came to the next building in the block about two hundred feet from the hotel. It was also a brick building that faced a side street with a large sidewalk and no entrance on the hotel side. Sleeping next to and against the wall of the building were children laying against each other. Boys and girls from 3 years to about 11 years old were snuggled together. There were no blankets or even cardboard boxes to protect them. He turned and went back the way he came with tears in his eyes.

Upon returning to the hotel, he asked the desk clerk who the children were, and the clerk said, "Saigon is full of street orphans. On Wednesday of each week, the Catholic Church opens their door to distribute clothing, and food that American churches have donated. The rest of the week, the children hang around the doors of restaurants to clean the plates of diners after they have finished eating. These children are gathering here for tomorrow the Church's doors will be opened for a few hours."

After hearing about the children, Gilbert began to think about ways he could help them. He was tired but did not want to return to his hotel so he went up to the roof where he could lookout over the city. As he stood there, he noticed three men sitting at one of the tables drinking beer. Their faces were only shadows from the flickering candle, but he could hear one of the men saying grace before they all lifted their glasses and drank. He could not help himself from staring. One of the men noticed and asked if he would like to join them. As they introduced themselves, Gilbert realized that they were Company Intelligence Officers, men with the same job as his on a break. He was still confused about hearing grace being given before drinking a beer, and when his turn came to give his name, stammered, "I'm Joe Kelly", and then said, "I'm Gilbert Moriggia." One of the men laughed and answered, "Don't worry. You are lucky. We have three names to remember." Then another man said, "Joe or Gilbert we are Mormons on leave from our provinces near Da Nang and are National Police Advisors on a few days R&R (Rest and Relaxation)."

"I didn't think Mormons drank beer?" Joe said. "We don't normally but the water in the towns is contaminated. Even the ice can kill you. We are from Provo, Utah, and our Bishop gave us a father and son talk before we left. You know. Beware of Eve's offering apples and don't drink the water. We wear garments under our clothing at home but were told to remove them so they would not become contaminated. Remember the lecture in Virginia where you were told to boil the water 15 minutes, put a drop of iodine in your canteen, anything to kill the bugs or one of the those little white pills in every glass of water before drinking? Three percent beer meets all the requirements, and it's an accepted drink by the Vietnamese. If you don't like the taste of beer try pouring it into a glass half-full of orange crush with one of the pills, and then drink it. That should work," one of the men answered.

While Gilbert was thinking about this, the man across the table spoke, "Just don't do anything that makes you stand out. You know, like saying grace over a bottle of beer."

Gilbert asked, "How do you know who are Viet Cong?" "It's easy. Look up at the sky. Its dark out and everyone can become a VC after dark. They know who we are, and we know that they know or something like that. When dawn comes the Dark Force disappears, and the Bright Force takes over. Those that believe in the Bright Force are Christians, and they will help you." Then he added, "I think the only way the Vietnamese will ever have a stable government is if the United States floods the land with missionaries and converts everyone." They all started laughing.

As Gilbert got up to leave, the first speaker spoke up while indicating the man next to him, "Take Dave here. Ever since the Bishop told us that there was a snake in this Garden of Eden he's been looking for an Eve to offer him an apple." Everyone laughed again, and then speaking in a serious voice one man said, "Beware of the women. They are the real power here. The average life expectancy of men in Vietnam is 40 years, and they really only think about money and women. The women control the money and the family while the men never get old enough to think about anything but girls."

It was around 2200 hours and Gilbert started worrying about returning to his hotel. As he walked to the entrance of the hotel, he noticed five black cars lined up waiting for their passengers. He spoke to the drivers who were sitting and visiting together. He asked, "How long will you have to wait for your passengers?" They all laughed and one answered, "I'll be here until daylight unless the General's secretary has another date lined up after he leaves." "Any chance I could get a ride back to the hotel Duc?" "Sure jump in, and I'll run you over to the hotel. Are you one of those Company Boys?" he asked. "Sure am. Don't you feel safer?" Gilbert asked. Everyone was laughing as they drove away.

The next morning Gilbert walked to the American Embassy. There he was briefed by the Communications Chief on the pseudonym or radio call signs for CIA Agents and evacuation procedures in case a green line (the perimeter of a compound or LZ) was overrun by the enemy. Nothing was said about the street orphans or how they could be helped. Gilbert could not forget and promised himself that he would fight the Dark Force forever.

He was assigned to the 4th Corp in Can Tho and told to spend the next three days reading Intelligence Reports. One report told how the Swift Boats, Rag Boats, River Patrol Boats, and Boston Whalers with an M-60 machine gun were patrolling the Mekong River trying to stop the enemy from using the river as a highway for supplying the VC. Swift boats (PCF) carried a crew of five with a Skipper, 50' long, 13.5' beam, two 50 Cal. Machine guns forward, and one aft along with an 81 mm mortar. It was powered with two engines of 480 hp each. Top speed 32 knots with a range of 250 miles.

Everyday he read the Province Reports of his predecessors while studying maps of his area of operations in the CIA reading room until the words started to blur. He kept looking for a pattern of the Viet Cong attacks. There had to be one but he had not found it. After eating supper in a nearby restaurant, he would walk the streets trying to think of someway he could help the street orphans. It was dark when he returned to his room, and all the children he saw while walking were curled up together next to buildings like sleeping puppies. He tried to count them but after a hundred he gave up, and walked back to the hotel choked up, almost crying to himself.

Tomorrow he would be driven to the airport, and Air America would fly him to his new post in Cao Lanh, Kien Phong Province on the Cambodian border. He still had not thought of a way to help the street orphans.

That night around two in the morning, the Eden Roc Hotel in downtown Saigon became the target of a mortar attack. Three rounds hit short by a hundred feet coming very close to the building next to the hotel. All three rounds hit on the sidewalk next to the wall amongst the sleeping children. They never woke up.

Now Gilbert knew where the war was, and it made him mad just thinking that the Dark Force could operate almost unopposed at night. He was eager and champing at the bit, planning on how he was going to capture the Communist Cell PRP leaders. Then they would be made to pay for their crimes against the children. It never even occurred to him that the Company had been trying to do this very thing for years. Every once in a while the CIA picks and recruits someone who does not know it cannot be done. They had found the man, and only had to give him the tools for the job.

In Vietnam, the drivers/mechanics for the American Embassy and the CIA were by government contract using Philippine drivers. These drivers were well trained, and this morning under a bright sun, Gilbert was being driven by a Philippine driver with all his baggage to the Saigon Tan Son Nhut Airport to be flown by a CIA Air America plane to Cao Lanh for his first assignment in the fight against the Dark Force. It was a single engine Swiss Porter plane painted in the colors of Air America Airlines, and he was its only passenger. Air America operated like a charter aircraft company when flying CIA missions. Each Company Agent had his own code that he used to schedule a flight, and Air America would bill the Company direct.

On the road to the airport, they had to cross a single railroad track. Traffic had stopped while a train with three freight cars was slowly backing into a siding to connect with two other wagons. Gilbert asked, "I didn't know the tracks had been repaired. Do you know where they run to?" "Last month the railroad was opened from Saigon to Loc Ninh at about the same time the Americans paved highway 13 to the Cambodian border. The tracks still need repairs to the north. There are rumors that Da Nang is next," the driver said.

Gilbert did not answer but just watched the train while thinking to himself. The intelligence reports he'd read in the CIA reading room did not mention anything about attacks on the railroad. Gilbert asked himself. Why would the Viet Cong let the railroad be repaired or highway 13 paved and never attacks the trucks or crews working on them? The North Vietnamese must be planning something big. He started making notes about how to get the VC to give him the information. While visiting the CIA offices on the third floor of the Embassy he had been told how the North Vietnamese Interrogated prisoners and it was giving him an idea. First, he had to capture a few Viet Cong to see if his idea of how to get them to talk and still stay within American's rules against torture would work.

After he arrived at his new station in Cao Lanh City he found that the quarters for the CIA Agent was a room with one wall part of a sandbagged bunker. It was the wall facing away from the prison that held the Viet Cong prisoners. The gun ports were placed in the wall five feet apart, and beside them were snappers (firing buttons) to fire the claymore mines that faced outwards toward the concertina wire surrounding the compound. The backside of each claymore mine was painted white so if they were attacked the defenders would know if the VC had somehow turned them to face the compound before firing them.

Just outside of the prison, there was a steel storage room about 10X20 feet close to the American advisors' quarters with one door and no windows. The room had been wired for electric lights and had a window type Sears' air conditioner installed in one wall. There were no prisoners allowed within the compound but the police still used the room so no one would know when they were questioning Viet Cong prisoners. Gilbert could walk from his quarters to the interrogation room staying within sight of the guard towers. Around the perimeter of the compound was concertina wire (barbed razor sharp rolls of wire), and guard towers located about two hundred feet apart.

His first night was typical of every night for the next four months. Gilbert was tired and had fallen into a deep sleep. The next day he wrote in his Journal; "I just got settled in my room next to the bunker entrance when we were mortared. I must have fallen asleep for I woke up on the run to my position at one of the gun ports in the bunker wall. The mosquitoes must be in league with the VC for they were waiting for me. After a couple of hours sitting in the dark being eaten alive by mosquitoes I'm beginning to think that incoming mortar rounds aren't that bad."

Gilbert never got used to spending his nights trying to get into the bunker just before the mortar rounds hit. One night he didn't make it and a mortar round went through the tile roof blowing him out of the door. The first aid station at MACV fixed him up, and the next day he started planning on how to capture a Viet Cong Communist and find out in advance when an attack would be made. Anything would be better than the mosquitoes.

A few days later the ARVIN PRU Chief and a Special Forces Sergeant Advisor captured an old man of around fifty who was a member of the Viet Cong carrying a rifle. The Province Intelligence Center (PIC) Interrogators had him in the room and were questioning him when Gilbert walked in with his Vietnamese Combat Interpreter Mr. Dien and his bodyguard a Buddhist Monk he just called "Monk". The Interpreter had a big thick mustache like Poncho Via wore in the western movies. When Dien learned that Poncho Via was a Mexican General he actually strutted as he walked. They entered the room wearing the black pajama uniform of the PRU Tiger Scouts and looked mean, fierce, and very serious.

Monk, the bodyguard spoke Vietnamese but no English and only a little French. He was a big man, and even with his shaved head, he was still taller than the other two men were. When he was wearing the black pajama PRU uniform he carried an M-16 slung over his right shoulder with the barrel pointing down and the sling towards the front. It was locked on full automatic fire, and he could swing the barrel forward, pointing it by turning his body while firing the weapon upside down. He always carried his begging bowl, knives, and throwing stars in a sack with his Buddhist Robe. When he was wearing his Buddhist orange robes his killing methods were with his hands and feet. This still made him the deadliest weapon in the room. He would take a position with his back to a wall and watch everyone including Americans. He saw no difference in who was the enemy. Gilbert had two commands that he could give by hand signal. A closed fist was to kill, and an open hand was to disable if possible. If Gilbert was attacked, there was nothing that would stop the mayhem.

The combat interpreter Dien with his dark handle bar dropping mustache was about the same size of Gilbert. It was easy for the Province Interrogation Center (PIC) interrogators to recognize who was the main man from just from the way the other two stood and moved with Gilbert. They never said a word but just stood and watched the interrogators ask questions. When they entered everyone in the room including the Province Interrogator Center (PIC) police thought that the prisoner's end had come.

The prisoner was sitting naked on a wooden stool in the middle of the room. His testicles had been wired to a field phone and when he didn't answer fast enough or with the right answer, he was given a jolt. The prisoner never even flinched or acted as if he felt the electrical charge. The Interrogators thought that the field phone coil was no good so they wired him up to a new phone but the results were the same. Gilbert told Dien to have the questioning halted. Gilbert said, "That man is so old and dried up he can't feel a thing."

The room had a small table with note pads and three chairs; also, the room was well lit. There was some trash and stored equipment along one end of the room but otherwise it was empty. Gilbert went over to the table and sat down facing the old man. The prisoner was held to the stool by plastic ties on his ankles with his hands tied behind his back.

Gilbert smiled at the old man while telling his Combat Interpreter Dien to remove the wires. Then he had everyone leave the building except the Interpreter. Still smiling he said. "Give him a drink of water but don't remove the straps. After he has drunk all the water he wants pour the rest of it over his head." "You know you can't be involved with torture. What are you going to do?" Dien asked. "I'm just trying to help him to see if he will talk to me. Move his stool so he is in front of the A/C and turn off all the lights except for two. Now turn up the A/C to high cool." Gilbert said. Then he walked out of the door followed by a puzzled Dien.

As Gilbert left, the interrogation room to find a restaurant that served cold beer and lunch he ordered his Combat Interpreter Dien to see that no one went into the building or bothered the prisoner until he returned. The Province Interrogation Center (PIC) interrogators were still waiting outside the building two hours later. This time when Gilbert walked in with Dien and Monk the prisoner started shouting answers to the PIC Interrogator questions before Gilbert could even sit down. He talked and talked. Gilbert just sat at the table with his tape recorder on. Finely Gilbert told Dien to ask the man to stop for a minute so he could ask a question.

When the man stopped Gilbert said, "Ask him if he knows where the mortar rounds that are being used to hit our compound are stored?" The answer was yes. "Ask him if he will show me?" Again, the answer was yes. "Untie him and give him some pants."

As they left, the Interrogation Room Gilbert met his Special Forces Advisor of the PRU Tigers Scouts in the compound. "Get some of your PRU Tigers and come with me. We are going to get those mortar shells that the Viet Cong have been hiding to fire at our compound. Maybe I'll get a little sleep tonight." Gilbert said.

"What the hell just happened?" Dien asked. Gilbert laughed and said, "Just look at our prisoner, he's at least fifty years old, dried up like a prune, and has never been cold before," and then continued, "His skin is so dry he never felt the electrical shocks, and even though that Sears A/C never got it colder than 70 degrees he couldn't stand it. Besides he knew I was trying to help him."

A few evenings later Gilbert again met with his Special Forces Advisor of the PRU Tiger Scouts. Both had been parachute jumpers at one time in their military careers and were becoming friends. The US Army military advisors had a small clubroom where they could relax. Colonel Joe Callahan, the US Army Senior Advisor, suggested that they all go over to the club so Gilbert could get better acquainted with the men he would be working with. Everyone had a beer in their hands, and Gilbert thought that this wasn't going to be a bad job after all. As the Colonel called for a toast the first mortar round hit. The VC were firing "HE" rounds this time which were much bigger than the rounds they had found, and the Viet Cong seemed bent on making it a miserable night. It is surprising how close you can get to know someone you only met an hour ago sitting together in a bunker drinking beer. Even the mosquitoes seemed excited to have the extra company.

As the new man, Gilbert the CIA Interrogator was assigned three jobs. The first was checking on the guards at night to be sure they were awake; another was head of the

Revolutionary Development Cadre, and the Officer In Charge (OIC) of the Static Census Grievances Committee. Gilbert had 720 men on his payroll that he personally paid once a month. Besides paying them, he had to keep them supplied with arms, ammunition, and bury them when they were killed. Gilbert had a few ghosts on the payroll but not enough to cause a ripple. Mostly it occurred when a man was killed his pay went on until he could not stand muster at the end of the month.

When the Viet Cong Communist killed one of his PRU Tiger Scouts he paid the family one year's pay, furnished a South Vietnamese flag for the coffin, and recommended the awarded of the Vietnamese Cross of Gallantry Medal for the fallen soldier.

He soon learned that the PRU Tiger Scouts priorities were money first, family second, and off in the distance somewhere Country. Whoever paid the most was who won their loyalty for the day. Today was today, and death had no tomorrow. What they wanted was a box to be buried in and to feel no pain. Gilbert was in luck for his new boss was a Company man, Caesar J. Civitella, who had been in the OSS during WW-II, and knew how to give a man the room to accomplish his mission. Gilbert's first priority was to learn how to work and move with the PRU Tiger Scouts to areas where they could capture VC. He desperately needed information on where the Viet Cong were operating, stockpiling supplies, and the targets they were planning on attacking.

One of Gilbert's first requests was to ask his boss for permission to contact the Tactical Operations Command

(TOC) and request US Navy Swift Boats to transport his PRU Tiger Scouts on night snatch and grab raids. This was given and followed by permission to also use US Army helicopters and trucks.

Gilbert had captured one Communist PRP Cell member when he first arrived in Kien Phong Province but none since then. Gilbert came to the conclusion that there was a mole or spy in the Tactical Operations Center in Saigon. (Later they found that this was true.) The snatch and grab raid worked because no one but Gilbert knew about the raid, where, or when it was to take place plus it all occurred within an hour. The only problem was he never got to conduct the interrogation of the prisoner. The Vietnamese National Interrogation Center (NIC) always transferred the high profile prisoners to Saigon for interrogation. The NIC Interrogators used the same impact torture techniques as the North Vietnamese Communist did in the Hanoi Hilton prison with the same minimal results.

With this success under his belt, Gilbert started developing a plan that seemed to be working. One thing that still had to be accomplished was to move the PRU Tiger Scouts closer to Cambodia where they could capture prisoners for interrogation without the enemy learning of his plans. Gilbert never told anyone in advance of where or when his Province Recon Units (PRU) where going. He would request pick up by the Navy at a LZ (landing zone) for an exact time and date but the Skipper of the Swift Boat would only be told the destination after the PRU Tiger Scouts were aboard. This same system also worked when using Army helicopters or trucks. No one but Gilbert knew the target, the time, or location of the raid until Gilbert informed the Navy Seal or Special Forces Advisor team leader. After the mission was over Gilbert, would read the after action reports, interrogate the prisoners, and then file his report through his supervisor to CIA Ambassador William (Bill) Colby. His rank of Ambassador was equal to the US State Department Ambassador, and he was in charge of all the CIA operations in Vietnam.

After Gilbert was assigned to the Ben Tre City, whenever he was given an order to remove someone Gilbert was never given a time or how to accomplish the mission. When it was impossible to capture an enemy, and they had to be removed for the safety of the Vietnamese people or American troops there was a sniper team available. A US Army Captain Barker, call sign Sandman, was assigned to support the Special National Police Squad. This was a last resort something like the police SWAT teams in the United States.

Gilbert started targeting Viet Cong in the outlying area of the Kien Hoa Province so he could get a better idea of where the VC were concentrated. There was a small island in the province where his informers had been reporting seeing a lot of fishing boats visiting. There were no crops being grown in the area so he asked the Navy to have a Swift Boat crew take his PRU Chief in at night to survey the island. The next day his PRU Chief reported that, it appeared to be a way station or supply point for the Viet Cong. Gilbert just thanked him and put a white pin in his map on the wall. The white pins designated points that had been visited and no VC action found. Gilbert hoped that anyone seeing the map would not know that he had discovered a Viet Cong supply base.

A few days later Gilbert radioed the Tactical Operations Center (TOC) and asked for a Swift Boat to give him an orientation ride. He informed them that the passengers would be a bodyguard, an interpreter, and a few police escorts. He did not specify a time but only that he wanted to see the area at night. He even told them if they did not have a boat available, another night would do. The Communist Viet Cong had begun to notice that Gilbert was behind most of their losses and were tracking his every move, but still had not realized how dangerous he really was. Gilbert had decided to raid the island but did not want anyone to even guess that he was going to hit the area personally, as he hoped to capture some important Communist leaders.

Ten days after his ARVIN PRU Chief had located the VC supply island Gilbert boarded a Navy Swift Boat with his Combat Interpreter Dien, bodyguard Monk, and a National Police Field Force Team. There wasn't space on the deck for another armed man. If there had been Gilbert would have had another police officer join them. The Navy Skipper asked, "I thought this was just an orientation ride?" Gilbert laughed and answered, "It is." Then he pulled out a chart and said, "No lights. Take us to this island and land here as he pointed at the map. Just drift in and don't fire any weapons unless you are attacked."

"This is supposed to be an orientation not a combat mission. Three of my crew have never been on the river before," the Skipper said. Gilbert once again laughed and said, "This is the police's first time too. I'm taking them out so they can get some experience in the field and gain some confidence. No smoking and no lights or sound. Turn your radios down. Let's just sneak in and out on the QT," Gilbert answered.

As they approached the island, the Skipper let both engines idle while drifting into the riverbank where Gilbert and his bodyguard Monk jumped ashore. They were followed by the White Mice (police) and Dien the Combat Interpreter. Gilbert spotted a large hutch (bamboo house) that looked too clean and orderly for a farmhouse in the center of the island. In it, they found a large radio that looked out of place and upon removing the back cover plate discovered that it was also a transmitter. Later when it was checked out by the Company's technicians in Saigon they were able to determine the frequency that the North Vietnamese Communist were using for issuing instructions to the Viet Cong.

While Gilbert was searching the hutch, the Police had found the supply dump and the Viet Cong. Small arms fire broke out, and the police panicked running back to Gilbert wanting to know what to do. It was now turning into a full-blown firefight, and retreat to the Swift Boat was out of the question. Those new Navy Sailors would not be able to tell one Vietnamese from another. Gilbert opened fire and had the most aggressive police officers move out to circle for an attack from the flank. Gilbert had told the police officers who went on the raid that the first men to get to the VC could keep anything they had on them. This was an important tool to motivate the police and the Province Recon Unit Tiger Scouts, but best of all it always worked.

The police were able to kill five Viet Cong and capture a few weapons, but no prisoners. The captured radio made the raid more than a success, for now the CIA could listen to North Vietnamese instructions to the Viet Cong. As they got ready for the return to the Navy Swift Boat Gilbert remembered that the Skipper had told him this was also an orientation cruise for his crew.

With the noise of a firefight followed by deep silence, this was no time to sneak up on a green Navy Swift Boat Crew. Gilbert told Dien to follow the last police officer, and then have everyone to begin singing. He started with "Working On The Rail Road" and everyone joined in. Now those Sailors had never heard Vietnamese and American trying to sing together before, but with all the laughing and singing, they were glad to welcome them aboard.

Gilbert received the South Vietnamese Cross of Gallantry with Bronze Star for his singing, and the North Vietnamese placed a bounty on his head. Even Colby had a few nice things to say. He really liked the radio.

One mission was given to Gilbert where this procedure was not used. Saigon Central Command had learned that some American Pilots were being held prisoner in tiger cages at a village near the Cambodian Border. The Tactical Operations Center (TOC) sent orders to Gilbert informing him that three Army Slicks (UH-1) helicopters would pick him and his PRU Tiger Scouts up for a rescue attempt at 1400 hours the following day. Gilbert sent for his Special Forces Advisor to the PRU Tiger Scouts and said, "I'll be joining you with my Vietnamese Combat Interpreter and my Cambodian Body Guard on this mission. The Tactical

Operations Command (TOC) has done everything except broadcast the mission on Saigon Radio. The enemy must know we are coming. I want to find out when and how the Viet Cong learned about our rescue attempt. Find someplace where the prisoners can be interrogated in the field, and burn the buildings when we leave." There were three American pilot prisoners moved north by the North Vietnamese that morning presumably heading to the Hanoi Hilton. The enemy had received word by a VC courier around 2200 hours the previous day but waited until 0800 hours that day before leaving the village.

The day following the unsuccessful rescue attempt Gilbert decided that the PRU Tiger Scouts should be moved closer to the border so the next time the Viet Cong would not have time to move their American prisoners. The Special Forces Sergeant Advisor and ARVIN PRU Tiger Scouts Chief had two Boston Whalers assigned for the trip up the Mekong River to look for a new home to house some of the Revolutionary Development Cadre (RDC) and PRU Tiger Scouts.

There were two Boston Whalers each with a crew of two, each driven by two 40 HP Johnson outboard motors, and each carrying an M-60 machine gun. Gilbert's boat had himself, his Cambodian Body Guard Monk, Combat Interpreter Dien, and four PRU Tiger Scouts while the other boat carried the boat crew, the Special Forces PRU Chief, another Combat Interpreter with a squad of six PRU Tiger Scouts. It was a beautiful warm morning to be on the river but upon leaving the town limits, they were in Viet Cong country or a Free Fire Zone, and anyone they met would be the enemy.

As they neared the Cambodian Border suddenly, the boats started taking small arms fire. The Skippers driving the boats headed them toward the riverbank at full throttle where the enemy was laying in the reeds firing their weapons. All of the PRU Tiger Scouts were returning fire, and when the boats M-60's swept through the VC they dropped like blown over bamboo reeds. The M-60 fire continued racking the riverbank and bamboo until the enemy fire stopped. Then the Skippers of the Boston Whalers turned and continued up stream. As Gilbert reloaded his M-16, he could see three bodies lying just out of the water but it was not a place they wanted to land. Those Johnson outboards could really move a boat, and they were long gone before the enemy could regroup. A week later when they returned the bodies were gone so there must have been more Viet Cong waiting to see if they would come ashore and walk into an ambush.

One day during the recon Gilbert located an old French mansion that had large rooms with over twelve foot high ceilings that could house the entire Revolutionary Development Cadre near a small village on the Cambodian border. The village was protected by a South Vietnamese Army (ARVIN) detachment commanded by a Captain Thoung. They accepted his invitation to stay for the night.

Gilbert decided to have a weapons check of the village police unit and found that most of the hand grenades were missing from their boxes. Gilbert asked the police unit leader of the village what had happened to the grenades, and was told that they were used when the village had been attacked a few days ago. Actually, the village fishermen would use hand grenades to catch fish. They would pull the pin and throw a grenade into the water where the fish were. After it went off the dead fish would float to the surface, and the fishermen would just gather them up.

Gilbert asked, "Show me the graves?" There was none. All the police officers of the village had gathered around listening and watching Gilbert. He picked up a grenade and calling over to the largest policeman watching said, "Throw this over the concertina wire surrounding the police compound." He left the pin in, and a good thing too for the man could not even throw it far enough to reach the wire. Then Gilbert said, "If the grenades are not in the boxes when we leave in the morning all the policemen will be drafted into the Army." Then he added that the Captain would be in command. Captain Thoung's face lighted up with a smile as if he had just won the national lottery. The next morning all the hand grenades were back in the boxes and everyone except the Captain seemed pleased with Gilberts visit.

After they finished picking the site for the Revolutionary Development Cadre's camp, they returned to Cao Lanh. A few days later, the Special Forces Advisor for the ARVIN PRU Chief was sent on a mission to capture Viet Cong prisoners from the area where they had visited. Gilbert was trying to obtain information on what the Viet Cong were planning about the pending move.

Gilbert's boss Caesar J. Civitella would not let any Viet Cong be interrogated in the Officer Special Assistant (OSA) compound even though most knew Gilbert was a

CIA Interrogator. The Special Forces Sergeant or the Navy Seal Chief would turn the prisoners over to the Province Interrogation Center (PIC), and Gilbert with his bodyguard, Monk and Combat Interpreter, Dien would drop in after the interrogation had started. This way he would be the good guy when he stopped the torture, and used his methods. The Viet Cong men always gave him the information he asked for but the Viet Cong women were another thing. They were the hardest to turn into police informers but when one did, it broke the back of their VC Communist cell.

One night Gilbert sent his Special Forces Chief with a squad of PRU Tiger Scouts on a night recon a few miles up the Mekong River using Navy Swift Boats to drop off his men and to return for them after the snatch and grab mission was over. They returned from the mission late that night, and put the prisoner they had captured into the bunker next to Gilbert's room. The Viet Cong had his feet and hands tied with nothing on but cloth tied around his privets to keep his testicles from bouncing when he ran. The mosquitoes had been waiting for Gilbert but seemed just as happy to feed on a VC wearing nothing but a lion cloth.

The next morning when Gilbert was brought in to interrogate the prisoner, all he wanted to do was talk if they would just let him scratch the mosquito bites. He agreed to lead them to a secret Viet Cong Camp outside of Cao Lanh with the PRU Tiger Scouts who knew how to look for Viet Cong bobby traps. It took two trucks to carry all the PRU Tiger Scouts with Gilbert and his Special Forces PRU Chief. Asking for the trucks must have alerted the spy in the

Saigon Tactical Operations Center (TOC) that something was about to happen. The Viet Cong did not know where but could guess when. As they moved up to the enemy camp one of the PRU Tiger Scouts accidentally set off a booby trap that contained a hand grenade, and the man fell against one of the trees, the trees were covered with ants, and when the man fell against the tree, they started falling from the branches. The more ants that fell on the PRU Tiger Scouts the more they ran into trees trying to get away, the more ants fell, until it was raining ants.

The Viet Cong heard the racket and fled. So far, not a shot had been fired, but when the grenade exploded, the VC informer panicked and ran. He fell into one of the punjy traps the enemy had placed throughout their camp. His screams just caused the Viet Cong to run faster, and with the rain of ants, there was no chance to capture any prisoners. Gilbert walked with his Combat Interpreter over to the hole where the screams were coming from and looked into the pit. There the VC informer lay with one leg impaled on a wooden spit of bamboo. The Interpreter asked Gilbert, "Should I shoot him now or drive a stick into his other leg?" All the PRU Tiger Scouts including Gilbert were dressed in their black pajama uniforms, and until Gilbert answered in Vietnamese any VC around must have thought it was the ARVIN Army who had found them. Gilbert said, "Don't shoot. Maybe I can help him." Then he spoke to the VC in the pit asking, "Do you know where the camp records are?" He said, "Yes". "If you will show them to me I will see that you are taken to an American hospital." The VC agreed between his screams of pain.

Two of the PRU Tiger Scouts then got down into the pit and lifted the man off the punji stick after giving him a shot of morphine to stop the screaming. He led them to the documents, which Gilbert took back to Police Headquarters where police intelligent officers could go over them. The documents included a complete map of 3F Army Camp and B-43 a Special Forces camp with the dates when they were going to be attacked. Best of all it had a document that indicated the S-2 in the Coa Lanh Tactical Operation Command (TOC) was a Viet Cong who had informed them about the forth-coming raid. When the Army went to arrest him he put on Buddhist Monks robes, and it turned into a political problem.

Gilbert read all the after action reports on each snatch and grab raid that the PRU Tiger Scouts went on. There was one report written by a Swift Boat Skipper that became the key to a plan he had been working on. This Skipper wrote in his after action report that he had taken one of his gunner's mates to the US Army's Third Surgical Division hospital at Dong Tan. While there, he observed and graphically wrote about a Tiger Scout who was being operated on by American Surgeons attempting to save his life. What he was doing in an operating room in combat clothing he never said but wrote that he was so confused and scared that he sat down on the operating room's floor feeling sick. It must have impressed those Army Doctors to see a Navy LTjg that afraid.

The LTjg wrote, "It seemed absurd---a man dying in his own country. I wanted to cry but I thought that I could not let myself and so tears just welled up in my eyelids. Now

I wonder why I didn't and I'm sorry." The LTjg did not seem to know that all PRU Tiger Scout were Viet Cong who had become police informers, and when the Viet Cong captured one they skinned him alive. The LTjg wrote, " I could see his neck bleeding, his head was arched back, his eyes half open, dazed searching for something...his left hand was wrapped in gauze... a pool of blood had gathered on the table below the green army stretcher on which he lay... everywhere there was blood pouring out of him. My stomach began to twist and sweat poured out of me."

As Gilbert read the Navy Skipper's words, it came to him as if out of the blue. Here was a man who did not believe that he had been saved by the sacrifice of God's son Jesus Christ on a Roman Cross. He had been turned to help the Dark Force, and while he may not have wanted to help the Americans, his actions that day showed Gilbert that the idea of fear could be as affective as the physical torture itself.

Gilbert went to the prison and using his newfound knowledge started interrogating Viet Cong who had not talked before. The idea was simple in reality. He promised to enlist them into the PRU Tiger Scouts and pay them as ARVIN Army Captains if they would give him the information he wanted or if they did not he would post their names on the prison bulletin board as being a candidate for membership in the local Province Recon Unit PRU. He would have them taken for a ride around the city in the back of a truck with PRU Tiger Scouts, and upon return give them presents in front of the other prisoners while thanking them. This procedure was always done after they had captured other

VC who were then placed in the prison. Even the new prisoners thought that it was the Viet Cong prisoner that Gilbert wanted information from who had pointed them out. They had a choice of being killed by the other Viet Cong prisoners for helping the PRU or really joining and being paid. It always worked.

There were no threats or torture used but only promises that Gilbert could and did keep. The Navy LTjg's fear helped cause the immediate death of over a hundred and twelve (112) Viet Cong Communist by the Sandman and Vietnamese Special National Police Squad (SWAT Team) over the next few days. Gilbert would identify the VC and then the police (SWAT Team) would take over. He was finely able to get a night's sleep without fighting mosquitoes in a bunker all night.

In his Journal, he wrote that he had recruited or turned eleven (11) girls who were working for the Viet Cong into informers for the National Special Police. Four (4) of them had provided information that provide directions for the US Army's artillery to shell VC targets and for the USAF to bomb North Vietnamese bases of the Province Sapper (terrorist) Battalion, the City Sapper (terrorist) Company, the Viet Cong Headquarters of the Current Affairs Province Committee, and the VC Military Proselyte (Recruiting) Province Section.

Gilbert had also been able to identify the following working Viet Cong:

7   Sapper (terrorist) Cell Leaders (City Level).
25 Sapper (terrorist) Members (City Level).

11 Military Intelligence Agents (Province, City, & District Levels).
4 Military Proselytizing (recruiting) Cadre (Province, City, & District).
1 Counter Intelligence Cell Leader (Province Level).
35 Commo-Liaison Agents (Region, Province, District, & Village Levels).
1 Province High School Teacher.
1 Village Education Cadre.
3 City Association Members.
6 Village Woman Association Cell Leaders
10 Village Woman Association Cell Members.
6 Village Nurses
1 Village Midwife
1 Front Line Laborer.

Total 112

Ben Tre City, Kiem Hoa Province is where the National Liberation Communist Party started the Viet Cong movement. They were well organized and used sappers (terrorists) to attack the Vietnamese citizens and government offices. It was the most dangerous city in Vietnam and the Dark Force was in complete control at night. Gilbert used his newfound knowledge of how to turn a Viet Cong into a police informer. Then with the help of the Sandman and the National Special Police (SWAT) Team, they were able to begin destroying the enemy's organization.

When Bill Colby heard of Gilbert's success, he sent Gage McAffe, Special Assistant to the Ambassador, to Ben Tre City to observe one of Gilbert's Interrogations. The

Ambassador also reviewed the records of the women and girls that had been recruited to help the National Police in capturing Viet Cong.

Gilbert's report was like a road map or blue print for the end of the Communist take over of South Vietnam. It was based upon Gilbert's observations, data from his Journal, and information from the Kien Hoa Province Senor Advisor US State Department's Ambassador I.E. Katzabue, his interpreter Pham Van Dien, and Captain Geoffrey T. Barker, US Army, Provincial Reconnaissance Unit Advisor.

He further attributed his first successful intelligence operation of turning a Viet Cong female into a police informer to careful planning, practical application of interrogations, and the concerted efforts of these men coupled with the cooperation of the National Police Special Branch, the Interrogators at the Kien Hoa Provincial Interrogation Center, the Province Recon Units, the Province Chief, the Province Security Committee, a local judge, and a US Army Medic.

Prior to the inception of Gilbert's methods of interrogation, the Special National Police had been unsuccessful in their attempts to obtain information from arrested female sappers (terrorist). Gilbert suggested to them that this failure could be caused by the impact torture method of interrogation. During this period, arrests of Viet Cong were mostly accomplished by military and paramilitary operations such as roadblocks and sweeps (searches), but the VC women remained true to the Dark Force and had never been turned or became police informers.

The first attempt to turn a Viet Cong female suspect began with the capture of a maid believed to have planted a bomb in the house of her employer, an ARVIN Lieutenant. Following her arrest, she was questioned by the police and the Province Intelligence Center interrogators with negative results even though physical coercion was used. Gilbert upon hearing about the arrest conferred with the police, and they agreed that he could question her. Using his methods and with acts of kindness the suspect implicated herself. She was subsequently recruited to work for the police as an informer. Having been successful in this instance Gilbert and Dien, his interpreter, decided to try and recruit more females as police informers.

During the following months the number of VC women informers increased to eighteen (18) varying in age from fifteen (15) to sixty-four (64).

With help from the Viet Cong, female informers, the National Police were able to capture more than one hundred fifty active Viet Cong. Many of those captured belonged to the province sapper (terrorist) unit, the armed security unit, military intelligence section, propaganda section, and there were also many VC communication agents of all levels from hamlets to province.

The capture of the VC communication and liaison agents critically disrupted the terrorist missions of the Viet Cong organizations causing a considerable decrease in Viet Cong terrorist attacks in Ben Tre City and the Kein Hoa Province. The VC Sapper (terrorist) Battalion D 570 whose primary mission was terrorism in Ben Tre City and the VC Armed

Security Company C 280 whose mission was kidnapping and assassination were destroyed.

As a result of Gilbert's work in turning VC men into PRU Tiger Scouts and VC girls into police informers the Viet Cong hostilities and attacks declined to almost none. People were able to again carry on normal activities in spite of the war. Through Gilbert's efforts, the first girl informer's prison sentence was commuted, and she was released. Nghi continued working for Gilbert and the PRU Tiger Scouts by accompanying them on covert operations. She would also ride around the city on a Honda 50 motorbike spotting VC for the PRU Tiger Scouts, which resulted in the capture of one female member of the military intelligence cadre of Chau Binh Village, Giong Trom District and another female commo-liaison agent from the same village in Kien Hoa Province. Two weeks later, she spotted a province level military intelligence agent who was then captured by the PRU Tiger Scouts.

To help make Gilberts program be successful in turning the Viet Cong women into police informers there was additional funds needed for the purchase of supplies and gifts. As there was no provision for these types of expenses, Gilbert paid for them out of his own funds. The total costs for the conversion of turning eighteen (18) female VC into police informers came to $150 US dollars. The number of American soldiers lives saved cannot be measured in dollars.

Gilbert's report also stressed the need for a rehabilitation program, which would be needed to preclude the girls from returning to the Viet Cong when released from police custody. The interrogation techniques for women were also much different than for the men. With the men, it was mostly money and discipline. For women the techniques were verbal subterfuge, psychological duress, sympathetic approaches, and by performing acts of human decency.

Gilbert established a system where the National Police or Province Interrogation Center would inform him when they had a likely prospect. The next four females called to his attention were also successfully recruited as results of his methods. Gilbert's Journal listed the first four females who became police informers and what was done to maintain their loyalty.

1. Nguyen Thi Nghi, (age 18), aka Anh Nghi, aka Tran Thi Lanh. She was born in 1951 in Chau Hoa Village, Giong Trom District, Kien Hoa Province. Nghi was of the Cao Dai religion. A graduate of junior high school after which she joined the Viet Cong. Her position with the VC was as a cell member and leader of a Sapper (terrorist) unit. During February 1969, she was hired as a maid for Nguyen Ngoc Tung an ARVN Lieutenant and S-4 Chief of Kien Hoa Province. She placed a bomb that destroyed the house on 1 September 1969 and was captured by the National Police. Nghi was subsequently locked up in a cell at the Province Interrogation Center (PIC) while waiting to undergo questioning.

Interrogation of Nghi by the National Police continued for three days during which times physical coercion was applied but failed to extract any information. Then the Province Interrogation Center (PIC) Interrogators questioned her for three more days with the same results. Failing to elect any information, they allowed Gilbert and his interpreter Dien to question her.

Nghi was blindfolded and was told that her visitors were an American Chieu Hoi (amnesty program) representative with his interpreter a Chieu Hoi cadre who had came to help her. Gilbert told her that the Province Interrogation Center was no place for an attractive girl, but it was necessary because she was a terrorist and some children had been injured, Dien asked her if she was beaten by the police, and she showed them some of the bruises. Gilbert told her that he did not approve of brutality and ordered the PIC medic to put something on the bruises to ease the pain. They were then escorted back to the interrogation room where Gilbert ordered them to remove the blindfold. Gilbert told her that she was very pretty and gave her a comb, an orange crush (soft drink), and some hard sweet candy but she still remained reticent. Dien told her that she was just being stubborn and did not want their help. Gilbert had been smiling as he gave her the comb and candy but now he looked hurt. He told her that because of her unwillingness to atone for her crime she was going to be executed by firing squad the next morning.

She became hysterical when told that children had been hurt and witnesses had testified that she had hidden the bomb. Gilbert told her that the hurt children would be invited to witness her death. Also that her ancestors would look unfavorably upon her because of the children unless she did something to atone for her gross misbehavior. She was given two hours to decide on a yes or no answer.

Having considered her dilemma she decided to cooperate with the police. Gilbert informed her that the decision had saved her life. She admitted to planting the bomb, and then submitted a list naming of thirty-five (35) members of Sapper (terrorist) Battalion number D 570 together with the names of other Viet Cong she knew.

Gilbert and Dien together with the police made plans to exploit her in the capture of Viet Cong. They would take her into Ben Tre City accompanied by a plain-clothes policeman to the bus station, water taxi station, the ID renewal station, the hospital and other places frequented by the VC. There were squads of police scattered around and PRU Tiger Scouts dressed as civilians close enough to protect her. When she pointed out a VC the PRU Unit would wait until after she left the area before making an arrest.

The first day of the operation passed with negative results. However, on the second day on the way to an assigned spotting point she pointed out two terrorists. On the third day she located three more and the next day another. For the next two months, she continued spotting for the police and was credited with assisting in the capture of twenty-two Viet Cong.

After every mission, spotting Viet Cong Gilbert would visit and congratulated her. He would give her small gifts such as candy, soap, shampoo, and other toilet articles. Nghi was moved from the Province Intelligence Center and given a room at the police station with an Army cot and mosquito netting. When she became ill, Gilbert had a US Army medic visit and treat her. Sometimes she would experience depression. Gilbert would give her money and take her to the market to buy fresh vegetables. She was also given reading material to inform her that the Vietnamese government had a definite interest in her and was trying to do something for the Vietnamese people.

On one of his visits, he noticed that Nghi was becoming very depressed. By this time there were three other girls living in a police barracks working as spotters. Gilbert informed them that the Province Chief and the Police Chief had asked the Judge to grant them a pardon. Then he gave them a ride through the city and down by the lake. It was an outing with just him, and no police. Dien and Monk followed in a second jeep and behind them, another jeep with PRU Tiger Scouts dressed in civilian cloths. It was not the girls they were guarding but Gilbert who was on the Viet Cong hit list.

After the ride, he stopped at a restaurant where Dien joined them for lunch. During lunch, Dien informed Gilbert that the Police Chief had asked that everyone return by 1400 hours. When the jeeps arrived at the police compound, there was a formation of police waiting for them. The Police Chief came out of his office and Dien led the girls to where they could stand in front of the police squad. The girls were nervous and looked worried. Gilbert just stood to the side and watched as the Police Chief thanked them for helping the people of Ben Tre City and handed each a paper releasing them from confinement.

In Nghi's behalf, the Province Chief personally requested that the Senior Advisor give her a job. It was a typist position, which was made possible because Dien paid for her tuition at a typing school. The talks and negotiations at the Paris

Peace Talks were going against the Americans and the rumors amongst the Vietnamese were that the United States was leaving. Many special programs were canceled and the typing job was gone. She was then hired as a maid for the PRU quarters with bonuses for captures of VC.

Nghi's blood has long ago soaked into the land. As a patriot of the Garden of Eden, this will be her final tribute to God and Country.

CHAPTER TWO

When the Americans left Vietnam, all the Vietnamese people who had helped defeat the Communist Dark Force were executed. Only those who had escaped with the Americans to the land of the Bright Force were able to live in freedom. The North Vietnamese Communist were very busy for the next few years hunting and executing all those who were strong enough to defeat them on the battle field. There were no trials, just find the people and kill them one by one. If there were children, they too had to die for they carried the genes of their parents, and would someday become defenders of the Bright Force. Once again, the streets of Ben Tre City ran red with the blood of its people.

The second terrorist that Gilbert captured was Le Ngoc Thuy, (age 16) aka Chau, a Sapper (terrorist) leader of the C 280 Armed Security Company. Just prior to her arrest, a fourteen (14) year old boy detonated a bomb that he was making in a house amongst the suburbs of Ben Tre City. He was wounded and taken to the hospital. Thuy went to visit him and was arrested by the National Police as a Viet Cong. She was born in 1953 at Nhuan Phu Tan Village, Don Nohn District, Kien Hoa Province. She is of the Buddhist religion and had attended three (3) years of primary school. During July 1969, she joined the Viet Cong and became a member of the sapper (terrorist) cell of Ben Tre City. In August, she also became a member of the Labor Youth Group, and in September 1969 was promoted to sapper cell leader. She was captured by the Vietnamese Police in October 1969.

Thuy's blood has also joined the tribute for the Garden of Eden.

The third terrorist captured was Nguyen Thi Bo, (age 21) aka Minh Khoa, aka Nguyen Thi Ut, whose name was already on the police black list. She was apprehended while trying to obtain a new I.D. Card. She was born in 1948 at Da Phouc Hoi Village, Mo Cay District, Kien Hoa Province. Bo is of the Buddhist religion and is able to read and write. In March 1969, she became a commo-liaison agent of the Standing Committee of the Province Communist Party Committee. In September 1969, she was admitted as a member of the Peoples Revolutionary Party (PRP). At the time of her capture, she held this position, and was a hard core Communist.

Bo's blood has also joined the tribute for the Garden of Eden.

The forth female terrorist that Gilbert turned into a police informer was Nguyen Thi Be Hai, (age 15) aka Be. She was captured by the police after being identified as a VC by Thuy. She was born in 1954 at Da Phuoc Hoi Village, Mo Cay District, Kien Hoa Province. Hai is of the Buddhist religion and has attended two (2) years of junior high school. Her last position with the Viet Cong was as a member of a suicide commando team, a section of C 280 Armed Security Company. She was arrested as a result of being pointed out by Thuy who took the same sapper (terrorist) training course with her. Investigation disclosed that her reason for joining the Viet Cong was that her father was killed by B-52 bombs and her mother by American artillery fire. Gilbert decided to recruit her as an informer.

Be's blood has also joined the tribute for the Garden of Eden.

Gilbert's interrogation methods were classified secret, reviewed, studied, changed, and became part of the CIA secret Dai Phong Program. Gilbert's basic idea was to turn the Viet Cong Communists into paid informants and enlist them into the Province Recon Unit (PRU) during the Interrogation. Gilbert just had not realized until he read the Navy Ltjg's written log that showed he was in sympathy with the communist and starting to help the Dark Force that it could happen without direct enemy contact.

When this Navy Swift Boat Skipper was turned, Gilbert suddenly realized it would not have happened if the man was a Christian. The Viet Cong were trained to fear torture and to love money. The only way to escape pain was by drugs or death. They had no loyalty to the Communist or to their country but for money they would serve the highest bidder. When the LTjg pledged himself, to a dieing PRU Tiger Scout who was a Viet Cong being paid by the Americans then Gilbert knew that anyone who didn't believe in God and Country could be turned. The Communist Viet Cong would skin alive any Vietnamese Christen or PRU Tiger Scout they could capture. Death was a release from the pain. Only true Christians who believed in God and Country were able to stand fast and remain true. This fact was shown to be true many times over by the US Navy pilots who had become prisoners of the North Vietnamese in the Hanoi Hilton.

When a new Navy LTjg volunteered to operate a Swift Boat (PCF) or PBR he wrote in his log or journal that he hoped to obtain relative safe costal patrols. He wrote that it was impossible to tell the Vietnamese men from the Viet Cong (VC) Communist enemy. The writings in his journal, and the letters he wrote home indicated that he was afraid when the Vietnamese looked at him as he walked down the city street of Da Nang, Vietnam.

His destination had to be the Da Nang Hotel's brothel as it had a large porch running the length of the hotel where a sailor could sit in the shade drinking beer while checking out the working girls (prostitutes). Besides there was no place else in Da Nang that a man in uniform who was afraid of the Vietnamese people could walk to. Another job that he was qualified for would have been the skipper of the boat ferrying Naval Personal to and from Tien Sha the Navy Camp at the foot of Monkey Mountain to the hotel. This was not to be, as Gilbert needed transportation for the PRU Tiger Scouts patrols from the mouths of the Mekong River on the South China Sea to the Cambodian border.

According to the LTjg's writings at the time, he had scant idea of what his men were supposed to do. He was also a very poor skipper as the boat's log shows that many times it got stuck in the mud or ran aground when the fighting started. Actually, he did not know anything about the Vietnamese people either or he would have realized that it was not the men but the women who controlled the countries economy and the families. They were the strength of their nation, and the real power behind the Viet Cong. The women even picked up restaurant checks, tipped the waiters, paid the family bills, and were actually the head of the average Vietnamese household. Only those women who followed the teaching of Jesus Christ took the man's name when they were married.

The Navy insisted on protecting their sailors from Venereal Diseases and enforced a health code on the Vietnamese Government. They did not have the money, time, or doctors to meet the Navy requirements so they improvised. The girls had to be 14 years old to work in the hotel or bars serving sailors. If a girl contacted a disease, she just disappeared. Moses had the same problem with the young Jews in the desert after they left Egypt and solved it in much the same way.

All the rooms in the hotel had running water. There was a tiled eight inch wide and four inches deep channel in the floor of the bathroom that ran through the walls to the next bathroom and then to the next room. Some of the sailors would blow up a condom like a balloon and float it in the channel that went from room to room. It would have a sign on it with a drawing titled Kilroy was here. The bed was of wood with a smelly pillow and a dirty thin blanket. There were no locks on the doors but it was actually a safe room as both the Viet Cong and the American fighting men used the hotel for the same reason at the same time.

Gilbert thought that the Navy LTjg was basically ineffective as a Skipper of a Swift Boat or River Power Boat, and had made little effort to understand his mission. The Vietnamese PRU Tiger Scouts needed transportation for their night snatch and grab raids along the Mekong River, and he had volunteered so the Navy assigned his boat along with others for the missions. In fact, the LTjg never reported that he had read or had been shown a Choi Hoa pamphlet published by the Viet Cong. It was a free pass for getting out of the US Navy. He never used it but after returning to the United States became a leader of America's Communist or anti-war movement. The LTjg was an educated and intelligent man who understood Navy regulations and used them. After submitting the paperwork recommending three Purple Hearts for himself, he was home free.

During his service in Vietnam while carrying Province Recon Units (PRU) Tiger Scouts to and from patrols the LTjg never discovered that he could have identified a communist Viet Cong by drawing the Christian sign of a fish in the sand. The enemy could not stand the sight and would have quickly shown him their dark side. The Christian Vietnamese people were his friend would have completed the drawing and greeted him as one of their own. There were many Vietnamese Christian people who were his friend if he would accept them.

During the war the Communist Viet Cong devised a peace sign to carry the anti-war movement to the United States, but when they saw the Catholic Cross with Christ crucified on it they did not see the cross but only how Christ's body hung upon it. They designed the cross in the shape of Christ's arms as he was crucified, and then enclosed it in an inverted circle so it hung upside down. South Vietnamese government Minister Mr. Cat Li tried to counter the sign by having it manufactured with one arm of the cross-broken to symbolize a broken peace. It never caught on in the US but the young LTjg became a prominent spokesman for Tom Hayden, and the Communist anti-war movement in the United States that used the symbol of the upside down Anti-Christ Cross as a peace symbol.

It is no wonder Gilbert did not ask for him personally to assist in the work of seeking out the enemy for the Navy LTjg had became a defender of the Dark Force. When someone is afraid of death to where he becomes ill just thinking of meeting his maker he becomes a tool for the Devil.

When Gilbert read, an after action report that the Navy LTjg had not fired his weapon in defense of his Swift Boat crew during a firefight he started to wonder why. Every sailor carried five hundred rounds of ammunition for his M-16, and the Navy was not short of ammunition. Gilbert was worried about the PRU Tiger Scouts night patrols, but had no direct control of the boats assigned to each mission. He just had to trust that the sailors would be able to protect themselves. To save ammunition in case it was needed to protect oneself is Un-American. To hide while the sailors fight to protect their captain is not the US Navy way.

The second man was an Army Chief Warrant Officer Aviator (CWO), Gentle Ben, who was 45 years old, and a citizen soldier who had served God and Country for over 26 years before arriving in Vietnam. He had volunteered so another Army Aviator could have a few more months home before he returned to the sometimes-unfriendly skies of Vietnam.

This man had served in WW-II as an Army pilot flying B-29 bombers and night fighters (P-61). During the Korean Police action, he flew the USAF F-82 and the F-86. The US Army sent him to Vietnam as an Army advisor for the South Vietnamese Army's Ch-47 helicopter company and approved his wife, Helen, to live with him. Upon his arrival in Vietnam during 1969, he requested assignment to "B" Company of the 228th Aviation Battalion, 1st Cavalry Division stationed on a Thai Firebase named Bear Cat. His command approved but continued his assignment for supporting the Vietnamese Province Recon Units (PRU).

This warrior was not just an Army Aviator but was trained for highly classified missions in the Far East, Korea, and Europe. He had served his country overseas in the US Army and in the USAF. Sometime during his training, he had read Haile Selassie's order to the Ethiopians when Mussolini invaded that country in 1935. "Everyone will be mobilized. Married men will take their wives to carry food and to cook. Those without wives will take any woman without a husband." He thought that this was a good order even if it was not in the US Army's rulebook.

Harry Alexandra, OSS, during WW-II had briefed Gentle Ben that he was not to follow an illegal order, and he was also taught a lot about how to fade out of the lime light. During the Vietnam War, Harry advised him to honor God and Country while serving in foreign lands with different cultures. Blend in, adhere to their laws, and adapt to the situation. Do not violate the laws of God or the United States.

The Vietnamese National Police had a 26-foot Australian trailer that a Province Recon Unit (PRU) had lived in behind its station. It had become a target for Viet Cong rockets at night so they gave it to Gentle Ben for his wife (Helen) to live in. She arrived in Vietnam two weeks later, and they moved the trailer onto the Thai Firebase behind the American PX (Post Exchange).

South Vietnam was a land much like the Garden of Eden or Shangri-la only it was under two governments. During the day under the light of the sun, it was the Bright Force supported by Americans soldiers, and at night it was controlled by the Dark Force supported by the Viet Cong Communist. A truck carrying rice on Hwy 1 from Saigon to Phnom Penh, Cambodia had to pay a tax to the government of the Bright Force and then stop at Go Dau Ha to pay the same tax to the communist government of the Dark Force.

Hwy 1 and the railroad ran side by side along the coast. They joined together near Nanning, China leading to Hanoi, Da Nang, and then to Saigon where the railroad turned west toward Cambodia. The railroad track paralleled Hwy 13 to Loc Ninh while Hwy 1 continued on to Phnom Penh, Cambodia where it joined Hwy 5 to Bangkok, Thailand. The Ho Chi Minh Trail joined Hwy 1 at Dong Hoi, North Vietnam crossing into Laos near Tchepone then south through Cambodia to join Hwy 13 at Loc Ninh with turn offs to Da Nang, Dak Bla, Piel Nong, and Kon Tum. It was the perfect and fastest transportation route to the rice basket of the Orient.

During the day, nothing moved on the Ho Chi Minh Trail, but at night, it turned into a major highway with bumper-to-bumper traffic. The trail looked like the Los Angles freeway during rush hour. When the sun went down the Dark Force came alive, and all the major roads in North Vietnam became clogged with traffic.

One night at about one in the morning, Gentle Ben was trying to contact a Province Recon Unit (PRU) Tiger Scout Squad that was operating just south of the trail when one of the gunners whispered asking over the Chinook's intercom, "Where are we?" They were at 2000 feet flying north of the intersection Hwy 13 and the Ho Chi Minh Trail. All the enemy trucks had their lights on and were jammed up at the bridge crossing into South Vietnam to connect with Hwy 13 and the railroad at Loc Ninh. It was the time of the monsoons, rain, thunderstorms, with the night sky being lit up with lighting bolts. On board the CH-47 Chinook helicopter everyone was whispering when Gentle Ben spoke up and said in a loud voice, "They can't hear us with all their truck engines' noise and the thunder. They can't see because of the rain so let's see how far the traffic is backed up and radio the Air Force."

After flying north along the trail for thirty minutes, the rain stopped. The lights from the trucks still stretched north as far as they could see as Gentle Ben turned and headed back into the rain shower they had flown out a few minutes ago. As they flew back into the storms the weather got worse and the helicopter's crew became very quiet as they headed toward Phuoc Vinh. Gentle Ben called Phuoc Vinh radio asking for permission to make an ILS approach, radioed, "This is C-47 Bravo 3467 requesting an ILS, estimating outer marker in 15, over?" "Roger 3467 cleared to outer marker 4000 feet contact Phuoc Vinh tower for approach clearance, Out".

In a Chinook, the intercom is hot, and when anyone spoke or coughed everyone could hear. Only the Aircraft

Commander or pilot could transmit outside of the helicopter but everyone could still hear them. The weather was getting worse, and the Chinook was shaking like a dog trying to get the water off its back. When Gentle Ben made the radio call for an instrument clearance the Chinook was at 2000 feet flying at 100 knots. No one had spoken or said anything to each other ever since they had gone to instruments flying. It was dark and the order was no smoking or lights. The pilot just pointed at the altimeter. Gentle Ben laughed and said over the intercom, "The Air Force thinks we are a USAF C-47, Puff the Magic Dragon. If they knew, we were an Army Chinook they wouldn't let us use the ILS. I've told them we are a C-47 at 4000 feet as that is the required altitude. Beside we are flying at 100 knots, and when we intercept the glide path, we will be right on the approach. Their GCA will never notice the difference."

The Flight engineer asked, "Wont they know we are a helicopter when they see us?" "They will never see us. The tower is reporting a 500 ft ceiling with ½-mile visibility. I will ask for a missed approach when the pilot can see the runway and turn over the helicopter to him. He will hover over to the Army log pad for hot refueling (refueling with the engines running), and then we'll scoot on down to Bear Cat staying VFR. I'll stay on the radio reporting the USAF C-47 flying toward Saigon. I will report VFR, and cancel the instrument clearance. No one will ever know we've been or how we did it. Tell anyone you want to but they will never believe it." One of the gunners answered, "Hell, I don't believe it now."

Whenever Gilbert needed movement or extraction by air for the PRU Tiger Scouts Gentle Ben would be there. The missions would be requested by Gilbert, and then a phone or radio call direct to Gentle Ben from the Special Operations Section in Saigon would give him the location and time for the pick-up. "B" Company furnished the helicopter and the crew but only Gentle Ben was given the mission information. Even he didn't know anything about the hard mission until the troops were aboard the Chinook.

Gentle Ben classified all insertions or extractions of Province Recon Units (PRU Tiger Scouts) as hard missions because the timing was critical, and someone was always shooting at the helicopter. Gentle Ben's flight log DA Form 759 records 737 combat hours flown. That is over thirty days of being continually shot at day and night. This was not like the Navy or Air Force because in the Army Gentle Ben could normally see or hear the person shooting at the helicopter. The Army Chief Warrant Officer Aviator's direct quote, "I hate it!"

The missions always started as a normal Company "B" supply flight for the 1st Cavalry firebases but when the time came for a pickup of ARVIN soldiers Gentle Ben would have his helicopter named the "City of Elkhart" to proceed to the pickup point for the mission. This seemed a problem with the Company "B" Commander as the helicopter was hit by enemy fire on four different occasions and in firefights one or more times a day at locations not actually in the company's area of operation. When the Chinook returned to the company at the end of the day there would be a Slick

UH-1 waiting to fly him for a debriefing in Saigon, and then back in time for dinner with his wife, Helen.

The policy in "B" Company was for all new Army Aviators in Vietnam to fly as pilots with an Aircraft Commander for the first six months in country. Gentle Ben just happened to be the only instrument rated helicopter pilot in the Battalion and an Aircraft Commander of a Chinook before he arrived. All the flight commanders in the company who had less than two weeks left on their Vietnam tour thought that it would be an easy mission if they flew with Gentle Ben. Most never lasted longer than one day. They had never been in two or three firefights in a single day before. They were not happy for someone who was supposed to be new in country taking the controls to drop into a jungle clearing to pickup ARVIN soldiers. On the third day he was in Vietnam, an Aircraft Commander who was a short timer (less than two weeks left in his tour) flew the helicopter back to the Company saying he had no orders to pickup Vietnamese soldiers.

It was the only time the Gentle Ben ever got back in time for lunch with Helen while flying with the 1st Cavalry Division. A Province Recon Unit (PRU) Tiger Squad was stranded waiting to be picked up. Gilbert was busy getting Slicks (UH-1) to make the pickup and hollering at the operations center in Saigon blaming them for the screw-up.

Upon landing, the Company Commander met them; before the Aircraft Commander could say anything, he and the crew were told to go into the debriefing room. Gentle Ben was asked to stay aboard the helicopter where the Company Commander handed him orders making him the Company Instrument Instructor Pilot (IP) and a Company "B" Aircraft Commander with a promise not to tell anyone. From that day on the company assigned him new black pilots to fly with his crew. Each black warrant officer aviator had to fly five missions with him before being assigned to other companies. In the period when he flew hard missions supporting Gilbert all the black pilots of the Battalion received more missions under combat fire than all the rest of the Chinook aviators put together and never knew why.

Gentle Ben was assigned to a flight under the command of Captain Jeffrey D. Hunger an artillery officer. On his first flight in Vietnam during the spring of 1969, they were flying near Bo Dop. It was an orientation flight of the 1st Cavalry Division area with no load aboard the helicopter when a radio call came from Saigon Special Operations requesting Gentle Ben to reply. Capt Hunger was in the Aircraft Commander's left seat, and he told Gentle Ben to answer the call. A new firebase near Bu Dop was under attack and being overrun by the enemy. The instructions were to pick up the last of the troopers who had held back the enemy while the Slicks (UH-1) got everyone else out. It was the time of year when the Monsoon Rains were coming down from the North sweeping through South Vietnam in the late afternoon and evenings. The rain was so heavy and the clouds so dark that it appeared to be a solid wall of darkness moving across the land.

The rains had come and time had run out. Saigon Special Operations wanted to know if Gentle Ben was near enough with a Chinook for one more lift. Maybe there was a chance to save the forty troopers still holding the firebase?

It was six in the afternoon, and they were flying in clear weather but could see a dark wall with lighting flashes over the boarder of Cambodia moving toward them. Captain Hunger had the controls and asked Gentle Ben to call the Red Hat radio operator on the firebase asking him to pop smoke. Five different smokes rose up all around where the firebase was supposed to be. They were at 4000 feet, and the smoke grenades showed two red, two white, and one green. The enemy was monitoring the radio calls and were trying to trap the Chinook. Gentle Ben pointed to the first red smoke off to their left. "That's our people," He said. The wall of water was so close that the helicopter was starting to bounce from the turbulence as Captain Hunger started down. "How do you know that's the correct smoke?" Captain Hunger asked. "I've got the Red Hat's radio tuned into the ILS needle, and the range bar indicates he's about a half a mile ahead," Gentle Ben answered.

The wall of water and the blackness was on top of them as Captain Hunger brought the helicopter to a hover just short of the green line. Visibility was only a few feet. Gentle Ben said, "Jeff stay away from the log pad. The Viet Cong have it zeroed in. Gunners do not fire your weapons unless we are being hit by small arms fire. No smoking, no lights." Captain Hunger turned the Chinook until the aft end was inside the firebase with the helicopter sitting on its belly astride concertina wire.

"Flight engineer ramp down but keep it level with the deck. Captain Hunger will lift the nose so the troops can board,"

Gentle Ben said. The rain was coming down in sheets with the only light coming from the lighting flashes, and the helicopter was rocking with each gust of wind. There were mortar rounds hitting all over the firebase and on top of the log pad area. "We've got a problem." Jeff said over the hot mike. "I agree those mortar rounds are getting dam close." Gentle Ben answered. "That's not the problem it's that I have never flown instruments, and there's no way I can get us out of here." Jeff said. Gentle Ben laughed and answered, "This is my business just give me the controls and follow my instructions. Aren't you an artillery officer?" Captain Hunger said he was and agreed to ask the artillery from the nearest firebase to adjust fire on his commands. He had fire laid down up to where the rounds were hitting directly in front and to each side of the helicopter.

The flight engineer called out over the intercom, "They are loading claymore mines and ammunition. Half of the troopers are going back for more. It's going to be a heavy load when we lift off." Capt Hunger answered, "Callout if we start taking any hits. Don't show any lights and don't fire unless you can see who is shooting." Then he added, "Those troopers don't want to face a bunch of claymore mines when they come back tomorrow. Let us know when the ramp is up."

It seemed like forever but only a few minutes had passed when Gentle Ben announced over the intercom, "Jeff, I've got a problem. You've got those artillery shells coming down so close I can't lift off into them." Jeff laughed and answered, "This is my business. You fly, tell me when, and I'll open us a path." Then he radioed the firebase firing the artillery and told them upon his command to stop firing, then wait ten minutes, advance range 300 feet, and fire for affect. He was not through giving his instruction when the Flight Engineer called out, "Ramp up."

As the Chinook started to lift off the concertina wire the shells stopped hitting, and Gentle Ben caused the helicopter to lift while climbing and turning until it was at 2000 feet on a heading toward Phouc Vinh radio for an ILS approach. The Flight Engineer called on the intercom saying, "Captain Hunger the Company Commander is a Captain with 40 troopers. Half of them are wounded. He wants to know where you are dropping them off." "Jeff, we've got to land for fuel at Phuoc Vinh, and I've already requested hot refueling." Gentle Ben said. "Tell the Captain I'm having a medical team meet us at the refueling area, and they can leave us there. Be sure they take all the mines and weapons with them," Jeff answered.

Gentle Ben wrote up a citation for the crew recommending a Silver Star for Captain Hunger and Bronze Stars for heroism for each of the crew a few days later. It was refused because the Company Commander told him that this was an orientation training flight, not a combat mission, and they were out of their assigned area. Gentle Ben had not submitted his own name for any award as he agreed that this was his first day in combat, and he had not volunteered. Only Captain Hunger and the crew had volunteered to follow him in the attempt to save the 1st Cavalry Division Troopers.

Along the Cambodian Border were strings of firebases all within artillery range of each other so they could provide fire support if they were attacked by the North Vietnamese Communist. One of these bases that Gentle Ben helped support was a firebase named Burt in Tay Ninh Province home of the 3rd Battalion, 25th Infantry Division. A job that the infantry grunts really disliked was the morning after fighting most of the night when they had to help bury the enemy dead and make a body count. Secretary McNamara tried to justify or prove that that the US military forces were wining the war by reporting a body count of enemy killed.

Before Gentle Ben arrived in country, the North Vietnamese attacked LZ Burt. This time the battle lasted from the 1st Jan to the 2nd January 1968. Sergeant David L. Walker was awarded the Bronze Star for heroism that day but then was given the job of counting the bodies. He counted 668 North Vietnamese killed, 23 American dead, and 156 wounded.

In 1960, Gentle Ben had attended the Air Force University as an USAF Major studying to become a nuclear weapons officer. His class had been briefed by an USAF Intelligent officer about France's fight with the North Vietnamese Communists. He insisted that if the French killed 10,000 of the enemy per month with conventional weapons that within ten years the Communists would overrun South Vietnam.

Using a bulldozer to dig and bury the enemy, it sometimes took the 3rd Battalion 25th Infantry Division a longer time than killing them. It only proved two things. Gilbert's plan of turning the Viet Cong so they would kill each other was the cheapest way, and the Mormon police advisors joke about flooding the country with Christian missioners to establish a stable government was not really a joke.

Gentle Ben was assigned to help Gilbert from 11 Jun 1969 to 1 December 1969. In these few months his DA Form 759 (flight record) records recorded that, he flew 737 combat hours. On the 8 November 1969, he was shot down at LZ Vivian in a firefight and was flown back to the Ft Rucker hospital on 5 February 1970.

All this time Gilbert was working trying to help the Vietnamese people, but he was not comfortable using only Province Recon Unit (PRU) Tiger Scouts whose loyalty started and ended with money. With Cao Lanh being on the border with Cambodia, Gilbert decided to recruit Cambodians to become guards for the Officer Special Assistant Compound (OSC). He personally supervised their training and even equipped them to go on snatch and grab raids in Cambodia to capture North Vietnamese soldiers to interrogate. To assure that they would be alert to any danger he authorized them to have their families live in a village area next to the concertina wire or green line surrounding the compound area (OSA). He also let his Combat Interpreter's family live within the compound.

Monk his bodyguard would always leave when Gilbert went into the compound or the officers' clubroom in the evening, and would join him in the morning as he left his apartment for breakfast. Once Gilbert asked Monk if he would like to sleep within the compound and was told that it was safer sleeping in a Buddhist Temple at night on his own sleeping mat. The incense smoke would keep the mosquitoes away, and there was no incoming mortar fire to keep him awake.

The first Cambodian Province Recon Unit (PRU) Tiger Scout platoon trained for snatch and grab raids was sent on a mission along the Ho Chi Min Trail to capture a North Vietnamese Army officer. They were all killed except for one man. Gilbert recruited another platoon of Cambodians and sent them out. Again only one man returned. It was the same man and after interrogation, he admitted to being paid for each man that he led into a North Vietnamese ambush.

During the interrogation, Gilbert promised that the prisoner would continue to receive ARVIN Captain's pay and be retrained if he told what had happened on the last two missions. He told how a North Vietnamese political affairs officer would gave him a location where the Cambodian PRU Tiger Scouts could be ambushed. After the fighting ended, every survivor alive was shot, and he was paid for each man killed. The US Army Special Forces Sergeant Advisor and the ARVIN Chief of the PRU Tiger Scouts was in the interrogation room and heard the man boast how good he was in leading the men into an ambush. The Advisor was mad and wanted to say something but Gilbert just smiled and said, "Not now."

After they were outside and while walking toward the small officers' club Gilbert said, "Sergeant, be sure everyone in your new cadre of Province Recon Unit (PRU) trainees know what happened here. Is there something you want to ask?" "Can you advance the repelling training for our trainees to take place tomorrow?" the Sergeant asked.

"I'll try and get the Army to send a UH-1 Slick helicopter for about two hours tomorrow starting at 1100 hours. If somehow they can't make it I'll let you know," Gilbert answered and then added, "I'll be sure everyone is waiting outside the club at 1130 hours. Better keep at least two armed guards awake watching to be sure our mole doesn't decide to leave."

The next day Gilbert visited Colonel Joe Callahan and asked him to inspect the PRU Trainees prior to the helicopter repelling training mission but never told him that he knew why PRU Tiger Scouts last two missions failed. The Colonel agreed to make the inspection, and made notes in his report that every man had his repelling belt fashioned correctly. Gilbert asked the Colonel if he had noticed that all the men wore new gloves. He just nodded his approval and joined Gilbert with the other advisors waiting for the officers' club door to be unlocked. "I wander what the hold up is?" the Colonel asked and then added, "I didn't even know they had locks on the door." Gilbert called out, "Look there's the helicopter. There are two ropes hanging below it. Maybe we'll get to see someone repel?" Suddenly a body came flying out of the door near one of the ropes. "He's not on the rope," someone called out. Another said, "The rope must have slipped out of the repelling ring."

The body hit about 500 feet in front of the compound gate. Gilbert and the others rushed up but the fall had killed the man. The officer who had said that it must have been the repelling ring that failed said, "He's got his gloves on but the ring is missing. It must have broken when he went out of the helicopter. Sorry, Gilbert, it's just one of those freak accidents."

Gilbert wrote in his personal journal; "There was no investigation of the accident as everyone saw it. No political hassle and no problem with the Vietnamese as the spy or mole was a Cambodian living in Vietnam. The CIA was billed for $350 dollars, and the cost of a coffin with a flag. Best of all the volunteering of young Cambodians to become guards has increased."

While Gilbert was in Cao Lanh, he would have hard candy mailed to him from the United States. He would carry the candy and give a piece to each Vietnamese child who washed their hands. They would run up and show him both hands. Gilbert would look at the little hands before placing a piece of candy in one of them. The children wanted two pieces of candy because they had two hands. If he only gave them one piece, they would run to the back of the line to start over showing him the other hand. He never forgot the street orphans of Saigon, and would do what ever he could to help all children.

The children were sick most of the time, and their parents did not know why. It was the water that made the adults and the children sick. When Gilbert used a Boston Whaler to travel the Mekong River they would go into canals through villages where people were bathing in the river and few feet upstream someone was using an outhouse that dumped directly into the canal. On both sides of the canal, people could be seen drinking and using the water to cook with. It was not only the Viet Cong that could kill you.

Gilbert wrote in his Journal, "Yesterday it rained and by evening the water was really coming down hard. I spent the day writing reports and did not make my usual trip into the village. No inspection of hands and no candy. Our compound is surrounded by concertina wire with claymore mines facing outwards just inside the wire. The gate had been closed and claymore mines placed across the entrance overlooked by a guard tower. It was not dark yet, and the guard informed us by field phone that the village children were all standing in the rain waiting to show me their hands. They were facing the wire and claymore mines while holding palm leaves over their heads to keep off the rain. Just as I looked out a bolt of lighting hit the fence, and all the claymore mines exploded.

I was blinded by the lighting bolt but ran out towards the gate while the rain was hitting my face like buckets of water. As my vision cleared. I could see all the children were still standing waiting for me with their palm leaves still held over their heads. As the guards and my officers helped clear a path, someone handed me my sack of candy. I gave each child two pieces and never even looked at their hands." Gilbert's last entry for that day read. "Thank you GOD."

One of the VC had been able to crawl along the wire in the rain, and with a long bamboo pole tipped the mines enough to aim them down so the blast and pellets went harmless into the earth. No one was hurt, and the accident forced the Dark Force to change its plan to use the storm to cover an attack on the compound. "It was God who protected the Children".

The next morning Gilbert had a wooden gage made that a guard could place over a claymore mine to determine its angle of fire. The officer who checked on the guards at night could also spot check the angle of the mines at the same time.

There was one more major change made. Gilbert would no longer give the kids candy in front of the compound. He would go into the village, and when he saw a child washing his or her hands, he would give them a piece of candy. In fact, he would never let the children line up or gather around him in crowds as the Viet Cong had put a price on his head, and he was now a prime target for terrorist. The award had started out as money, but now it included a hutch, food, and maybe even a bicycle.

Gilbert had a cistern built to catch and hold rainwater for his Tiger Scouts to bathe in but when they went home to their village they still used the Mekong River water for drinking. He remembered the three Mormon police advisors in Saigon telling him that it is the water that can kill you. When you drink beer, wash your teeth in beer, and cook with beer he could understand why they never wanted to drink beer again.

One morning he received a radio call from Cao Lanh Province Senor Advisor Colonel Joe Callahan, "I have Woody Hays, football coach, and Marty Karow, baseball coach, from Ohio State University in my office. They want to visit one of our outposts in Viet Cong territory to show some football and baseball movies. All of the Choppers (helicopters) are scheduled for the next few days. Will you take them on one of your boats to an outpost on the

Cambodian Border?" "I'll be glad to help. Do they know that this is enemy territory?" Gilbert asked. Then he added, "If they still want to go bring them over to the boat dock in thirty minutes. I'll meet them there and pick the outpost to visit after we leave the dock."

Gilbert told his Army Special Forces Sergeant Advisor/Chief of the PRU Tiger Scouts to pick four men and meet him at the dock. He didn't tell him the mission or even that they were carrying two passengers. When the Sergeant arrived with his men Gilbert pointed at one of the two Boston Whalers, with an M-60 machine gun, tied to the dock and said, "Sergeant have that boat crew prepare for a three hour cruise along the canals. Be sure no one leaves the docks until after we are gone."

As they were getting the boat ready, a jeep pulled up driven by Joe Callahan with two men and two large boxes about the size of large suitcases. Gilbert did not introduce the two men but asked the Sergeant to seat them one on each side between two PRU Tiger Scouts. As they pushed off from the pier Gilbert announced in a loud voice so all could hear, "We're going on a cruise of the canal toward the city and return by another route." As they slowly moved out of hearing, Gilbert introduced the Sergeant to Woody Hays and Marty Karow of Ohio State University.

"We're going to an outpost established in a village on the Cambodian border. It's about a thirty minutes run with the two outboards at full throttle. Our passengers will be showing football and baseball movies to the men at the outpost. We'll stay there about an hour, and then return before dark," Gilbert said.

The Sergeant just stared at him and then said, "You know this is crazy. We will be traveling through enemy territory in a free fire zone. There is no guarantee we won't be ambushed." Then he continued checking the M-60 making sure it was ready to fire at a touch on the trigger.

They arrived at the outpost during the noon period, and the Sergeant helped carry the boxes to the compound where the movies would be shown. The boat crew stayed with the Boston Whaler while Gilbert took the four PRU Tiger Scouts to a small house bar in a hutch with a palm-thatched roof. There was a small table in front where they sat drinking Ba Mi Ba in the Vietnamese manor with ice in it. As Gilbert stirred the ice in the beer with a small stick, he started to wander how they could be drinking beer with ice in a village with no electric near the Ho Chi Minh trail. There was no electric or refrigeration around for miles.

As he stirred the ice in the glass the bar owner came over and said, "Its good ice. It was delivered this morning." Gilbert suddenly lost his taste for beer and stood up heading for the compound to tell his passenger it was time to leave. As he stood up, he saw Woody walking toward the bar. It looked as if he was coming to get a cool drink. Gilbert met him half way steering him toward the boat. The Sergeant was walking with Marty carrying the other box. "Sergeant there is ice in the beer. Bring the men and let's leave." Gilbert said. All the Sergeant said, "Ice." He abruptly ordered the PRU Tiger Scouts to follow him and changed his path toward the boat.

The Tiger Scouts were not in a hurry but when they realized that Gilbert was pushing the boat off from the dock, and they would be left behind they jumped in. The Sergeant stood up and manned the M-60 while the two-crew members kept the boat at full throttle. With two 40 HP Outboards pushing the Boston Whaler it rushed down stream on the Mekong River. As they approached the canal turning into Cao Lanh the boat slowed to make the turn while the Sergeant sprayed the banks with M-60 machine gun fire. The enemy started firing back with AK and small arms fire. The crew went back to full throttle while the Province Recon Unit (PRU) Tiger Scouts joined the fight. They were through the ambush, and the PRU Tiger Scouts were all laughing.

No one aboard the Boston Whaler was wounded, and if the people of Ohio ever found out, they would know that these were two gutsy coaches but they knew that already. The pucker factor was pretty high that day though. Gilbert and his Special Forces Chief had another night of ducking incoming mortar rounds while Woody Hayes and Marty Karow were already back in Saigon having a cold beer without ice.

The next morning Gilbert had breakfast with his Advisor/ Chief of the PRU Tiger Scouts. They had become friends even though the head of his Tiger Scouts was Special Forces, and he was a retired USAF Sergeant. Each had saved the others life on raids into enemy territory. "Sergeant," Gilbert said, "I want to recruit more Cambodians, and have you train them. Today I'm going to visit the Cambodian village next to our compound, and start making some friends."

Some people may think that the Vietnamese and Cambodians looked alike, and some did but they were two different people. His Vietnamese PRU Tiger Scouts were Viet Cong

Communists who had hired out helping Americans to capture and identify VC terrorist. The Cambodians were Buddhists who wanted to fight the Communist. It was a big difference.

A few days later the Sergeant, Dien, Monk, and Gilbert drove a jeep into the small Cambodia village built up against the wire behind their compound. Gilbert took soap and hard candy as gifts for the children. His plan was to use the children to earn the trust of the people. They stopped in the center of the village and Gilbert started his routine of checking the children's hands. Those who washed them got a piece of candy. The Sergeant and the Interpreter stayed in the jeep and watched Gilbert work.

Gilbert noticed an old man setting in the shade of a tree watching them. His left foot was elevated and rested on a box. Gilbert bowed and greeted him in the Buddhist Manor with respect. Clasped hands as in prayer and a bow from the waist. The old man's foot had an old cut that was festering from infection and proud flesh. Gilbert called for his Interpreter to bring him his emergency medical bag. After giving the old man a shot of morphine to kill the pain, he gave also him a shot of Vitamin B-12.

All the children had gathered around watching with very serious looks as Gilbert cleaned his foot and cut away the dead skin. As he worked the old man would rub his head and stroke his arm while calling him some name Gilbert didn't understand. All the children would laugh and clap their hands. Every two or three days Gilbert would return, and the children would gather around laughing as the old man kept stroking him. After about three weeks, two young men came to the Compound and signed up to be OSA Compound guards, and the old man started pointing out the Viet Cong tax collectors and sappers (terrorists).

After two weeks of Gilbert, cleaning the wound the old man's foot had started healing, and he did not need the pain pills. Gilbert would just wash his foot while the children giggled. The Sergeant and the Interpreter would be talking to young men that met Gilbert's requirements for guards as he worked. He asked the Interpreter, "What is the old man saying when he is stroking my arms or patting me on the head?" "He is calling you his monkey" the Interpreter answered, "It's a Cambodian greeting of love."

A message from Bill Colby at MACV had been delivered to Gilbert that morning informing him that he was being assigned to a new location to start implement the CIA secret Dai Phong Program. What Colby did not tell him was that when the Program had been approved the US State Department had not thought that it would work and had argued for an agreement with the North Vietnamese at the Paris Peace Talks for the US Army to withdraw from South Vietnam. At this time, there was no longer a Communist organization of Viet Cong in Ben Tre City, Kien Hoa Province. All the Communists had surrendered or been killed, but the State Department Ambassador did not want to hear this. Colby told Gilbert at one of their meeting that the Ambassador had relayed this information to the Paris Peace Talks and that the North Vietnamese had assured them that this was not true.

Colby knew that a deal was on the table for the North Vietnamese to return the Navy pilots who were prisoners in the Hanoi Hilton but it would take time for the prisoners to be fed and to get their health back enough so that the American people would accept the Communist Lies. The agreement was for the prisoners to be released when the North Vietnamese found them well enough, and for those that couldn't be returned due to mental problems caused by years of torture or crippled so bad their injures would show were to be classified as missing in action (MIA). The military would continue their pay until their bodies were found or they were declared dead.

No one asked what would be done with these prisoners but only that they would be listed as MIA. The North Vietnamese had no mental facilities to treat these kinds of injuries so they treated them as they did the Vietnamese Christians and the PRU Tiger Scouts they captured. The citizens of the United States were treated with an avalanche of lies and told that no members of the Communist Central Committee had ever been captured. The American people were told that all American prisoners would be returned as soon as the United States left Vietnam. This was a lie as Gilbert had already identified and captured three Communist members of the Peoples Revolutionary Party (PRP). One was turned over to the National Intelligence Center and two over to his Province Intelligence Center (PIC) for him to interrogate. He was able to turn these two, and they helped provide the means to defeat the Viet Cong in their province.

Bill Colby must have known it was too late to stop the avalanche of Communist lies and propaganda that was turning the American people against fighting to save the Vietnamese people from the Dark Force. The Communist goal was to show the world that Americans could lose a war and be destroyed with out having to fight their Army, Navy, or Air Force. The Paris Peace Talks (Peace Accord) were part of their plan of defeating the United States and destroying the Vietnamese people.

Colby was a patriot of the highest order, He wanted Americans to someday learn that the Communists and the Dark Force could and must be defeated if America was to remain free. His plan was top secret and was known only to a few in the CIA but it was so simple that no one would know what had happened unless they studied, and understood that the Vietnam War was also a test of how to destroy the American's will to defend the Bright Force.

He decided to divide South Vietnam along the highway from Kon Tum (Kontum) where the Ho Chi Minh Trail joined Hwy 14 to Plei Ku (Pleiku), then along Hwy 7 to Cheo Reo, and then to Tuy Hoa. The division would only be noticed by the North Vietnamese because there would be no Communists Viet Cong in this zone.

It would also give the American's time to get all their people out of Vietnam, as the North Vietnamese Army would be so busy killing the teachers, and the professionals who had helped the Americans that it would take weeks before they even started to wonder how the Communists Viet Cong had been driven out of the area.

It would later become a road lined by Vietnamese dead for forty miles that marked the freedom route. He hoped that someday American historians would wonder how South Vietnam became divided by a Bright Force Zone where there were no Communists or Viet Cong. He hoped that when the Americans were again fighting the Dark Force they would understand that the Communists and the Terrorist could be defeated, and maybe learn how it had been done. He also believed that freedom must be fought for and protected at all times.

Gilbert first met Captain Geoffrey T. Barker, US Army when he led a counter attack against the enemy that helped save the lives of the American Embassy personal in Ben Tre City during an attack on that Embassy 7th & 8th June 1970. Gilbert had been wounded during the firefight but was still able to help USA General Charles Timmes to safety, and to organize the defense of the Embassy. To prevent the American people from knowing that the North Vietnamese could attack and penetrate the defenses of an US Embassy in 1970 the firefight was classified secret. This also included censoring all the medical records of the men wounded in the action and the American State Department denying that it had happened.

There were no purple hearts awarded or medals presented as these could also bring attention to what the North Vietnamese Army could do and would take attention away from the deals being made at the Paris Peace Talks (Peace Accord). When Staff Sergeant Barker was an advisor for the PRU Tiger Scouts General Timmes recommended him to be promoted from Staff Sergeant direct to 1st Lieutenant, and later to become an advisor for the PRU Tiger Scouts in three different Corps areas.

When the US Advisor/PRU Chief in Ben Tre City was killed (KIA), Barker was sent to Ben Tre City and assigned the

CIA Call sign of Sandman. He and Gilbert became friends working together on a secret CIA Dai Phong Program Test. Both had finely learned that it was the Vietnamese women who were the power in the Viet Cong. They concentrated their efforts on capturing VC girls and turning them into police informers. Once a girl identified a Viet Cong infiltrator, she would make a prearranged signal. Two PRU Tiger Scouts would follow the suspect until out of sight of the informer and then snatch them up. The girls never made a mistake by pointing out the wrong person. Every Viet Cong that was identified and captured was carrying weapons or explosives with them. The program was a complete success.

The Vietnamese Province Chief at Ben Tre City was always trying to obtain information on the inner working of the Province Recon Unit (PRU). He assigned Lieutenant Thanh as the ARVIN PRU Chief. This man was a typical ARVIN non-producer and always tried to pad the monthly report for enemy killed or wounded. He did not want to take second place to a VC turncoat and was against the program. He had been turned by the Dark Force and was keeping names with information so he could buy a position from the Communists if the Americans left Vietnam. There really was a snake in the Garden of Eden.

One month an anomaly occurred. Usually the PRU Chief tried to pad the monthly report but this time his report was one less than Captain Barker was reporting. Captain Barker doubled checked and confirmed that his tally was correct. He sent his interpreter to make some discrete inquires. They discovered that a female Viet Cong who was captured through the Dai Phung Program was in a Province Hospital following an interrogation by the PRU Chief. Captain Barker with his interpreter located the VC in the city hospital and convinced her to make a statement. With this information, he now had sufficient evidence to have the ARVIN PRU Chief fired. She also agreed to become a National Police informer and worked with the PRU Tiger Scouts.

Captain Barker wrote in his Journal on the 22 September 1970 that he felt a little conspicuous wearing a US Army uniform for the first time since he arrived in Vietnam. On Province Recon Unit (PRU) missions he wore sterile (no identification) tiger-stripes or the black pajama uniform of the PRU Tiger Scouts. He was at the Ton Son Nhut Airport seated on the freedom bird (airliner) waiting for the engines to be started. Just as he was getting used to the idea that the plane was really headed home two American men in dark suits complete with eyeshades (glasses) came down the aisle. They stopped at Captain Barker's seat and asked him to accompany them off the aircraft. They were armed and placed themselves so he had no option but to comply with their request. As they walked down the movable stairway, Captain Barker saw that they had already removed his luggage (parachute kit bag) from the aircraft, and it was on the tarmac. Securing his bag, they walked into the terminal where the suits produced MACV CID credentials and read him his rights. Captain Barker was told that he had been accused of torturing two prisoners who had been in his custody, and had murdered at least one of them.

They asked him to make a statement but he declined and requested more details. Captain Barker was told that on the 27 April 1969 he had taken two prisoners away from the Province Recon Unit (PRU) compound in Ben Tre City and never brought them back. One had been found tortured and killed outside of the city. Two witnesses had testified that they had asked him not to take the prisoners but that Barker had dismissed them. The second prisoner's body was never recovered.

Colonel Tex Hyslip, USMC was head of the PRU Program with its Headquarters located in Saigon. After Captain Barker's arrest, he met with him and recommended that Captain Barker make some kind of a statement. Colonel Hyslip explained, "This will become a Vietnamese Military Tribunal made up officers of the Buddhist, Cao Dia, Hoa Wao and other religions who do not place a hand on the bible or swear to tell the truth so help me God". He cautioned that it would be Captain Barker's word against theirs, and the hearing was on their turf. Captain Barker thanked him but still refused to say anything.

The Military Tribunal was held in the IV Corps area. The Vietnamese JAG (Law Officer) read the charges, and then requested a response. Again, Captain Barker declined exercising his right under Vietnamese law to have them prove the charges.

Captain Barker was not surprised when the former ARVIN PRU Chief walked in as the key witness. Everyone listened to this man's testimony about how he tried to persuade Captain Barker not to take the prisoners or to harm them. His testimony was corroborated by his cousin who was a supply-man with the PRU Tiger Scouts in Ben Tre City. They both added that Captain Barker was extremely drunk at the time and would probably claim he did not remember the incident.

Following this damning testimony of the two star-witnesses, Captain Barker was again asked to make a statement. This time Captain Barker advised the Tribunal that he had documentation that would prove the witnesses were lying. He reached into his wallet and pulled out his military ID card. He placed it upon the IAG the desk. Captain Barker then asked if the JAG (law officer) knew what this was. The JAG officer answered saying, "This is an American US Army military ID Card issued to Captain Barker the accused." When asked if there was anything significant about the card the JAG officer confirmed that it had his photograph, army service number, birth date, and had his signature on it. Everyone thought that Captain Barker was going to invoke his diplomatic immunity that all military personal had been given while serving in Vietnam. They were surprised when he asked the next question, "Was his birthday on the 27 April?" All the Tribunal members came alert as they tried to watch both him and the witnesses at the same time. This was the same day the witnesses had said Captain Barker killed a prisoner and would not remember what he was doing.

Captain Barker then apologized to the Tribunal for not making a statement before but that it was necessary to appear before them so that the witnesses could be exposed as Viet Cong supporters. This was a serious charge, and would become a life and death matter for someone. Captain Barker then reported that he and two other CIA employees had departed from Ben Tre City on an Air American flight to Can Tho on the 26 April 1969 to attend the change of command of the CIA Regional Officer in Charge. They stayed over an extra day to celebrate his birthday, and returned to Ben Tre City via Air America on the 28 April 1969.

After returning home, he was notified by the MACV Inspector General that the Military Tribunal had dismissed the charges. He also learned that the former PRU Chief had an unfortunate accident and did not make it back from his last operation with the Province Regional Forces.

While Captain Geoffrey T. Barker was on the freedom bird flying home to the United States, Gilbert was busy planning an ambush of a Viet Cong Sapper company (terrorist group) between the towns of Hong Ngu and Moc Hoa. His informers had provided information that these Viet Cong had been trained by the North Vietnam regular Army and were planning to infiltrate Saigon to attack President Nixon during his Vietnam visit. Gilbert's problem was that if he told anyone that the President was coming to Vietnam they would not believe him or if they did Gilbert would not be allowed to do anything or worse he might be suspected of being a threat to the President of the United States. It was having information that was so important that Gilbert could not tell anyone and yet he had to do something.

Gilbert did two things that worked but only because no one believed the information or would believe that he could stop the Viet Cong before they could attack the President. The intelligence was especially suspect because it was obtained from the Viet Cong prisoners who he was interrogating. Gilbert sent his PRU Tiger Scouts on a secret snatch and gab raid, and then he radioed the Saigon Special Operations Center (TOC), and asked for Gentle Ben to pickup a special sling load of claymore mines and ammunition at the Long Bien Ordinance Depot for a PRU Tiger Scouts' ambush.

The Special Forces Sergeant Advisor to the ARVIN PRU Chief and the PRU Tiger Scouts were to setup an ambush on a small dirt road between two towns. The small dirt road passed the east side of Saigon ending up where it joined Hwy 15 at the Bien Hoa Airport. Gilbert's instructions to the PRU Special Forces Advisor was that they were going to snatch and grab all Vietnamese who would be traveling on the road toward Saigon. Gilbert did not tell them that the group they were looking for would outnumber them two to one or that the Viet Cong had to be captured or killed or the President of the United States could be in great danger.

He also forgot to mention that the PRU Tiger Scouts could not carry enough ammunition and claymore mines for the ambush or that he was going to deliver what they needed by a CH-47 helicopter at the ambush site. Gilbert used his CIA ID Card for the authorization to make up a sling load of weapons and "C" rations that would be needed for the ambush to be successful. He was at the Army Helipad (Log Pad) waiting for Gentle Ben to arrive thinking that he might have to hold the road for as long as three days. Gilbert was wearing a PRU Tiger Scout sanitized uniform except that he had put on subdued 1st Lieutenant Bars along with crossed Infantry Rifles. His idea was to keep the raid as low keyed as possible just in case something went wrong.

As Gilbert waited at the Army heliport, he saw a 1st Cavalry Division Ch-47 coming in. It landed on a roll with its nose twenty feet in the air, and as it turned to face the sling load, its ramp came down so he could jump aboard. As Gilbert came aboard the helicopter the ramp closed up, and he saw the gunners at their posts with their M-60 machine guns pointed down towards the ground. Gilbert ran forward to stand between the two pilots, and as he looked back into the cargo hold, the flight engineer was already laying on his belly looking down through the center hole at the sling load. The helicopter's deck was level, and it was slowly moving forward where the loadmaster was waiting to hook up the sling's donut to the helicopter's cargo hook.

Gentle Ben had no idea where the destination was going to be but that was the normal situation when you flew for the PRU Tiger Scouts and Gilbert. Gilbert was standing between the two pilots' seats wearing the 1st Lieutenant Bars of an infantry officer. As he came aboard, he had noticed that everything was coated in red dust. The dust was the same color as the dirt near Ft Rucker, AL where they had trained. The crew were all wearing goggles that completely covered their eyes and were sealed to their red faces. Gilbert was standing under the forward transmission, and the noise made his teeth rattle. The man in the left seat had the controls and was maneuvering to pick up the load. The man in the right seat was writing on the windshield of the helicopter with a grease crayon. The writing appeared to be the time and a dash to indicate direction.

Gilbert wanted to tell Gentle Ben the destination and direction where they were heading but all he could see were two pilots who looked alike. Both were wearing two piece nomex flight suits, bullet board armor, nomex gloves, each had a helmet on with the visor raised, each had on the same type of flying goggles as the rest of the crew, each had on a shoulder hostler with a pistol, and both were completely covered in red dust. The pilot in the right seat grinned at Gilbert and pointed to a headset with a throat mike hanging on the back of his seat. He lifted his goggles letting them hang around his neck. He had white rings around both of his eyes and held out his hand saying, "I'm the Aircraft Commander, Lieutenant. Where are we headed?" Gilbert hadn't put the throat mike on but held it up to his neck with one mike on each side of his throat. Both of the pilots were using the same type of mike, and the noise of the transmissions did not affect his hearing.

He did not tell Gentle Ben that he was Gilbert, and Gentle Ben didn't seem to recognize him from the times they had met in Okinawa or France. There was no need to endanger him or involve him in this adventure of trying to protect the President. If he was successful, no one would even know what had happened except the Communists in Paris. Gilbert's information had came from someone in Paris who knew of the Presidents travel plans and then through a Viet Cong prisoner so it was hard to know who to trust.

Instead Gilbert answered the question by telling Gentle Ben to take a heading southwest toward the small town of Moc Hoa in the II Corps area and to maintain radio silence. Gentle Ben reach up and took the controls while announcing on the intercom to the crew that they were on a "no one needs to know mission" and at the same time he reached up to the radio controls turning them off. The other pilot was now grinning and took off his goggles hanging them around his neck. Gilbert found himself looking at a red dust covered man with eyes circled by black rings where the goggles had protected his face. Gentle Ben said Lieutenant meet George. This is his second day flying with "B" Company, and I'm getting him ready to be an Aircraft Commander.

"Why did you turn off the radios?" Gilbert asked. "You said radio silence, and I can't refuse an order from the Saigon Tactical Operations Command (TOC). Therefore, if I cannot hear it I can't refuse it. Besides, we have a transponder on this helicopter, and with it squawking any radar station can track us. Whenever I fly spook missions, the radios always seem to fail or develop trouble. Don't worry the crew chief knows how to fix them, and will soon have everything back to normal," Gentle Ben answered. Then he said, "George, take over and level off at 2000 feet." He reached up and using the grease crayon marked down the heading and time, they reached cruising altitude.

A short time later, they were approaching just to the east of Moc Hoa and Gilbert said, "follow the little dirt road leading out of town for about three miles, and we should see some of my soldiers." "Will they pop smoke or give us any directions?" Gentle Ben asked. "They are Vietnamese dressed as I am, and some may be in black pajama uniforms. They will not even look at the helicopter as we drop our load. I'll jump out off the back ramp after you clear the load." Gilbert answered. Then he said, "After I get off proceed to somewhere in the III Corps area before you fix your radios. We will find another way back to Ben Tre City."

George spoke up, "That road looks too small for us to land on." Don't worry just hold the helicopter steady and moving forward fast enough so the dust cloud will be just behind the Flight Engineer's hole in the cargo deck. When the Flight Engineer says stop just bring us to a gentle stop and don't change your altitude. He will lower the sling to the ground. Remember watch your instruments and keep level using the artificial horizon, use the needle ball and airspeed. I will be on the controls with you but don't pay any attention. You just fly your instruments and I'll fly mine. This way our load will stay steady and not start swinging. After we are clear of the load the Lieutenant will hang onto the hook, and the Flight Engineer will lower him to the ground. Everyone check that your goggles are tight to the face. Gunners keep your weapons pointing towards the ground. We don't want anyone to get nervous," Gentle Ben instructed.

"There's soldiers running into the trees about a click (a half mile) ahead," one of the gunners called out. George, start your approach from here and remember to level out just so our blades are not hitting any trees. Move forward until the Flight Engineer says, "Stop." Then Gentle Ben said, "Lieutenant, as soon as the Flight Engineer drops the load you go back and let him lower you to the ground." The Flight Engineer called out over the intercom, "Loads clear." Then the Gilbert called out, "these are North Vietnamese soldiers. They are not my Tiger Scouts." Gentle Ben with out a change in his voice ordered, "Gunners keep your weapons pointed down. Stand back from your hatch and tell me what you see. George, move us forward and keep the dust cloud just aft of the gunner's positions. After we are out of range for small arms fire climb up and turn to a down wind leg. Level off at 300 feet. We are going back for the load before the dust settles and the enemy discovers we aren't in it." Then he asked each gunner what he had seen. When Gilbert started to say something Gentle Ben just held up a hand indicting not now.

One gunner said that the soldiers he saw were hiding their faces away from the helicopter and the dust but that they were defiantly uniformed Vietnamese. The other gunner said that the men he saw had their backs toward the helicopter but he could see that they were carrying AK47 weapons. "George, keep it in tight and no base leg. Put us down right above the road and make lots of dust. Engineer get your Shepard's Hook ready to snatch the donut. (A Shepard's Hook is a hook on a long pole that a Flight Engineer can use to connect up a load while laying down looking through the center hole.) Gunners track the sling load, and if we start taking fire try to destroy it. If we are lucky the load will explode and give us time to get out of here," Gentle Ben said.

"The loads hooked," the Flight Engineer called out.

The Chinook climbed straight ahead and Gentle Ben asked, "Lieutenant, what do you want to do now?" "Climb to five hundred feet and follow the road. My people must be close," Gilbert answered. About a quarter mile ahead the road made a slow turn to the east, and then back towards Hong Ngu. Just as the road turned, Gilbert called out, "there they are. Put the load down. I'll get off with it." "Lieutenant, that was a North Vietnamese ambush set up to hit your patrol. Are still going to establish an ambush two miles from them on the same trail?" Gentle Ben asked. "I know it doesn't make sense but it's important that no one goes down the road towards Saigon. The North Vietnamese were just trying to keep anyone from blocking the trail but my Tiger Scouts were already in position so they have failed. I don't think they know I've got the trail closed. You can still help me by making a radio call to a FAC ship (USAF spotter plane for fighter-bombers) and target the North Vietnamese ambush. A few USAF 500 lbs bombs should take their mind off of what I'm doing," Gilbert said.

"If you can get your radio repaired by the time you get back to the III Corps area it will help us," Gilbert said. As the Chinook began climbing, Gentle Ben said, "George, level off at 4000 feet and head north. We will refuel at the Quan Loi Airport." "That will take us across Cambodia is that Ok?" George asked. "Push up the speed to 100 knots while I see if I can contact a FAC ship. Tell me if you see a sign that reads Cambodia," Gentle Ben said while laughing. One of the Gunners asked on the intercom, "Do you think that was a real Lieutenant, Sir?"

CHAPTER THREE

Two days after President Nixon left Vietnam Gilbert received a phone call from William Colby's secretary in Saigon who told him that Mr. Colby would like for him to stop at his office on the way to his next assignment. She also said that Mr. Colby wanted him to know that the President was pleased that there was no attempt by the VC to interfere with his recent visit. It was Bill Colby way of letting him know that he knew about his ambush of the Viet Cong sappers. This was a strange war where the men fighting it could not be recognized or given credit for what they were doing as it might make some politician look bad or the people of America to continue standing with the Vietnamese people against the Dark Force.

"Can you hear me?" Gilbert asked the Province Recon Unit (PRU) Tiger Scout sent to pick him up at the dirt airstrip where an Air America Porter aircraft had dropped him off. The sound of the aircraft engine revving up for takeoff made it almost impossible to hear what the driver, dressed in the black pajama uniform of a PRU, was saying as he approached. "Welcome to Tuy Hoa." The airstrip was nothing more than part of a dirt road where Air America landed with freight or passengers and was not connected to the Tuy Hoa USAF Airbase.

This was the first town that Bill Colby had sent Gilbert to since he proved the success of the CIA secret Di Phong Program in the Kien Hoa Province. He was sorry that he had to leave Ben Tre City, but it was getting a little boring not having incoming mortar rounds or rockets keeping everyone awake at night. This was a chance to prove that the CIA pacification program would work wherever the Dark Force terrorists were attacking with rockets, bombs, mortars, terrorist, and their most successful weapon, an avalanche of lies to destroy the moral of the people.

It was only a 15 Minute Drive from the airstrip to his quarters which turned out to be a room in an old French Mansion with its high ceilings that made Gilbert think he was sleeping in a tennis court. The house faced the South China Sea with a driveway from Beach Road to the front entrance. The quarters for the Special Forces Sergeant Advisor and the PRU Tiger Scouts were in a "L" shaped building along the north and west side of the compound facing the house. There was a storage shed on the northeast corner for a jeep and boat. A Concertina Wire fence completely circled the Compound with its gate opening onto the dirt road and across it a white sandy beach.

After putting his bags in the room, Gilbert went immediately to the office of the ARVIN Province Chief in the MACV Compound to meet with him and his staff. The Headquarters building not only held the Province Chief's offices but all the government offices. It was a sit-down meeting that went very well. He was told that every night the Dark Forces took over control of the city with their mortar and rocket attacks. They all wanted to talk to someone who came from a Vietnamese city where the people weren't always afraid and hung on to his every word as he told about sleeping with out being awakened by rockets or mortar attacks at night. Gilbert wrote in his Journal that it was strange living in a country were the dinner conversation was about incoming rockets or mortar fire instead of rain.

The briefing included information on all the ARVIN soldiers in the area, the National Police Compound, and the Province Recon Units (PRU) Tiger Scouts under the guidance of the US Army Special Forces Command. He was also informed that there was a Korean Marine Division located just south of the Tuy Hoa Airbase who provided the largest military ground combat presence in the area. There was also an Army Ordnance Facility located south of the USAF Airbase beside Hwy 1 before the Hwy passed the Korean Marine White Horse Division. It wasn't as large the Ordnance Facility at Long Bien near Bien Hoa, Saigon, but its mission was the same; providing ammunition and bombs for the military in the area.

Tuy Hoa Air Base was only a five minutes flight by helicopter to the south from the PRU Tiger Scouts Compound but an hour drive by vehicle. The Vietnamese Hwy 1 and the railroad track ran just to the north of the USAF Air Base and the Korean Marine Compound. It was the main route from Guangzhu, Nanning China, Hanoi North, Vietnam, continuing south to Da Nang, Tuy Hoa, and to Saigon South, Vietnam. Due to Viet Cong ambushes in the area it was a very dangerous road to travel on even for short distances.

After Gilbert had settled into his new quarters he started learning his way around the city. It was a routine that he went through with his bodyguard upon being assigned to a new area. This time Monk, his bodyguard, who was also a Buddhist Priest, asked him if he could have a day off. It seemed that a truck had run off the highway in the small village of Phu Khe and had hit a Buddhist Spirit House. The business that kept the house as a home for the spirits had sent word that they needed a Buddhist Priest to appease the spirits and asked for Monk to return. It meant a large feast, and the foods that the spirits didn't eat would be given to the business' guests and the Buddhist Priest to feast upon. Gilbert approved and decided he would use the day to visit the USAF Air Base and Korean Compound.

It was a bright and sunny day around 1000 hours with little traffic on the highway. Gilbert planned on visiting the Korean Compound for an hour just to look it over, and then drive to the USAF Base for lunch at the Officers Club located on the beach of the South China Sea. He wasn't planning on meeting anyone but just to look the compounds over. He had done this in the city of Tuy Hoa the day before with his bodyguard Monk and his new interpreter. He wouldn't need his bodyguard or interpreter today as he would be on the Air Base or with the Koreans, and the short drive between the bases would be made in his jeep at his usual full speed or flat out. He had been taught to drive fast and defensively by the Special Forces anti terrorist team incase he was driving a VIP trying to avoid kidnapers. Gilbert figures that if this was the way to save a very important person it was the best way to protect himself from the Viet Cong when he didn't have guards along.

He had to slow down as he moved amongst people pulling carts and walking across the railroad bridge over the Song Ba River. The Highway Bridge had been blown up by the Viet Cong during Tet of 1968 and work on rebuilding it hadn't started yet. Having to slow down to cross the bridge didn't bother Gilbert but it did alert the VC that he was alone and would be returning over the same route in a few hours. It gave them time to set up an ambush between the Korean Compound and the USAF Airbase.

When Gilbert stopped at the gate of the Korean's firebase he asked for permission to drive around the compound. The guards couldn't understand him but as he waited they found a marine sergeant who spoke English. After showing him his identification Gilbert asked, "May I drive through your compound to observe your marine combat training?" The sergeant went into the guard shack and used the phone. He then came out and asked Gilbert to get into the right front seat while he drove. Another guard carrying an M-16 climbed into the back, and off they went.

The first thing Gilbert noticed was that the normal concertina wire around the compound had another single strand of wire a foot off the ground and fifty feet further out. Gilbert asked, "What is the single wire used for?" "It is the notification wire. Anything crossing it day or night is killed. If it's crossed it is dead. It doesn't matter if it's a dog, Viet Cong, person, bird, or a marine. At noon every day a patrol goes out and picks up everything killed in the last twenty-four hours," answered the Sergeant. The jeep stopped and the sergeant pointed out a patrol waking twenty abreast looking at the ground. Ahead of them were three soldiers also studying the ground, and then came soldiers carrying sacks for anything the searches found. Following them were five soldiers pulling and sweeping the ground with branches. The sergeant explained, "The first three men are looking for tracks in the dirt. If any are found the soldiers guarding that area will be punished if the thing making the tracks hasn't been killed. The men sweeping the ground are preparing it for the next shift of soldiers going on guard duty." "What kind of punishment will the soldiers receive if any tracks are found?" Gilbert asked. "Every man in the company and the officers will lose one months pay and six months without a pass or R&R. Its never happened," the driver answered.

The jeep started up again, and they drove toward the canal that boarded the southeast side of the compound. As they approached the canal the sergeant stopped the jeep to let an officer with a sergeant and six marines leading a prisoner pass. The officer was leading the squad walking toward a clear area facing the canal away from the barracks. The prisoner had a hood over his head and a rope around his neck held by a sergeant. Gilbert asked who the prisoner was and his guide answered, "He's a marine who had raped a Vietnamese girl and was going to be shot. Would you like to watch?"

As they drove back to the gate Gilbert asked why they didn't give the man a dishonorable discharge and send him back to Korea. The Sergeant said it would disgrace his family, and they wouldn't allow him to stay in their house. He had also dishonored the marines, and the Colonel would lose face. Besides if they shot him before returning him home his family would receive all of his pay for years to come, and he will go back as a hero. All the Korean veterans who return home dead or alive are treated as heroes forever.

It was a lesson that Gilbert never forgot for when he returned home a year later in a medical back brace and body cast no one wanted to hear about Vietnam. There were no parades or honors for Vietnam Veterans and least of all for a civilian who worked for his country trying to help save the lives of the Vietnamese people. Just knowing that a person was a Contract Intelligence GS-11 Agent of the CIA made him an untouchable. In the two years after the Americans left Vietnam there were more Cambodian and South Vietnamese civilians killed than during the period from 1962 to 1973. There were even some people elected to public office in the United States who claimed to be leaders of the Vietnamese Anti-War (Anti-Christ) movement.

The Korean Sergeant stopped at the gate, and Gilbert changed into the driver's seat. As he entered Hwy 1 he turned right heading towards Tuy Hoa. As he picked up speed the road passed over a built up area between two small hills. It was about noon, and he was thinking about stopping for lunch at the Officers' Club on the Airbase. Suddenly he heard small arms fire and saw the rounds hitting the road just in front of the jeep. He went to full throttle and started swerving to confuse the enemy's fire. Just at that moment a small boy with three cows started driving them across the road. To avoid them Gilbert went into the ditch beside the road and the vehicle turned over. After rolling over once it stopped upside down in the ditch.

There wasn't a sound as Gilbert peeked out from under the vehicle but he couldn't see anyone so he used his radio and while lying under the jeep called for back up. He heard a clang and rattle from a bullet hitting the jeep as a sniper fired to see if he was still there. There wasn't much room between the jeep, and the bottom of the ditch. When he crawled on his belly trying to look out from under the jeep his ass hit the steering wheel. The Province Recon Unit (PRU) Chief who answered his call for help radioed, "We're ten minutes away. Stay cool."

The previous day Gilbert had his PRU Advisor assign PRU Tiger Scouts to patrol the highway from Tuy An north of the city to Phu Khe south of the Korean Compound. There were two PRU patrols working the highway, one to the south and one to the north. This was the patrols second day and the first radio call they had received asking for help. It took 15 minutes before they arrived, and during this time Gilbert had to lay in the ditch with his 9 mm pistol in his hand waiting for the Viet Cong to come down to see if they had got him. If they did came down and looked under the jeep they were going to get a surprise for he wasn't about to become a prisoner while he could fight back.

The PRU Tiger Scouts arrived loaded for bear but all they could see was a jeep upside down in a ditch. The PRU Chief was laughing as he called down to the overturned jeep and told Gilbert that he could come out now as they were there to protect him. The PRU Chief was teasing him, and the men were laughing. Just then the Viet Cong opened up with small arms automatic fire. Gilbert had to push the PRU Chief off of himself as there was no more room under the jeep or in the ditch. All the soldiers were in the ditch with him and returning fire. Without a word being said two PRU Tiger Scouts crawled southward in the ditch and two others in the opposite direction. They then crossed the road to flank the ambushers. One of the Tiger Scouts in the ditch fired a M79 round into the hill where the enemy's fire was coming from, and then came a brief burst of fire from the flankers. The PRU Tiger Scouts came walking back to the ditch to help Gilbert and their PRU Chief climb back upon the road.

One of the flankers was showing the other Tiger Scouts a watch he had found and another was showing them some other trinket. Gilbert didn't even display any interest as he had a rule that whoever killed a Viet Cong in a firefight got to keep any personal items the enemy may have on them. As he had no use for dead bodies, and no one was paid for killing a VC there was no rule about burying any dead enemy soldiers.

Gilbert agreed with Gentle Ben that making a body count report was a stupid waste of time but regulations were regulations, and burying a dead enemy with a bull dozer or without a box was an insult to their enemies' ancestors. After a firefight was ended Gilbert would have the PRU Tiger Scouts bring in their dead and wounded. The wounded would be placed in an American Army or USAF hospital, and the CIA would pay the dead man's family a years pay; furnish a coffin, flag, and a Vietnamese medal. Gilbert would caulk the seams in the coffin as a sign of respect and honor to the soldier's family. The enemy's dead were left where they had fallen, and the next day the bodies would be gone. The North Vietnamese and the Viet Cong did the same thing that Gilbert did for his dead. Only the winners had wounded.

Everyone was talking a mile a minute, and no one seemed worried about standing in the middle of the road while traffic was waiting for someone to flag them on. The boy with the cows had been watching from the side of the road as his cows grazed on the thick grass. The PRU Chief had the truck turned around and helped Gilbert into the passenger seat while he took over the drivers position making the driver get in the back. As the traffic started moving no one drove or followed too close to the truck. Only the boy with the cows waved good-by as they passed him.

Every morning Gilbert would turn left out of his driveway, and drive about a mile before coming to the turnoff that brought him to the National Police Compound where his office was located. The compound also had the police station and the Province Interrogation Center (PIC) in it. All the buildings were within walking distance of his new office. During the day the only thing Gilbert found strange about the people was when he was talking to the women in the Tuy Hoa street market the first question they always asked was, "How old are you?" Followed by, "How much money do you make?"

Gentle Ben told Gilbert that when he was asked these questions he would take off his hat or cap and show them his white hair. This always got a pat on the arm, and if they were Buddhist they would ask to rub his belly for luck. When they got around to asking about money he told them that Helen, his wife, was very rich and made as much money as their president. Then they would ask where his wife lived, and he would show them her picture in front of their trailer at Bear Cat, Vietnam. This stopped the questions.

Beach Road continued west past the turnoff to the National Police Compound for another ½ mile before coming to the single railroad track and then Highway 1. Turning south or to the left on the highway, and after crossing a bridge over the Song Ba River the road would pass the turn off for the gate to the USAF Airbase. Tuy Hoa Airbase extended east of the tracks and highway to the South China Sea. The railroad track and highway continued south pass the Korean Whitehorse Marine Division's Compound on there way to Saigon.

Tuy Hoa had another prominent landmark that Gentle Ben (Gilbert's friendly Army Aviator) always looked for before landing on an LZ near a city or town. It had a Catholic Church with a steeple and an annex on the corner of Beach Road and Hwy 1 just east of the railroad tracks. This was a signpost that he used to inform his gunners that they would not recon by fire upon landing. This rule didn't apply at night when the Dark Force had control of the city.

Gilbert had worked with Cambodians, Vietnamese, but never with Koreans up until now. After he left the meeting with the Province Chief he went to the Province Interrogation Center (PIC) and told them he would like to see all of their people in a sit down situation. He didn't want a lot of flags flying or speeches of about what they had done in the past, but more of a meeting on how to capture Viet Cong officials and sappers (terrorist) to find out what the enemy was planning.

Gilbert noticed that they were not a proud bunch as he put it, but were just sitting around because they had nothing to do. They had no prisoners in custody and hadn't since the Korean Whitehorse Marine Division had moved into the province. They acted as if there would never be any more Communists captured. It was an atmosphere of lost hope.

The grass had grown so high it reached eye level around the PIC office and interrogation building. They must have known Gilbert was coming as everyone was in uniform. Some of the uniforms looked so new they didn't have a crease in them. The Vietnamese really take care of the person who pays them, and Bill Colby had made sure that everyone knew that Gilbert not only paid them but controlled how much they were paid. This was not the Korean way or the US military way but it sure made Gilbert's word law with the Vietnamese.

If that wasn't enough to get their attention there was his bodyguard "Monk" looking and acting like he would enjoy tearing someone apart. When Air America flew Gilbert to Tuy Hoa City he brought his bodyguard with him but had left his combat interpreter Dien in Ben Tre City as he was needed there. He told the Province Interrogation Center (PIC) people that he would meet with them the next day, and the grass must be cut around the interrogation building. He insisted that the building should also be painted white inside and out before his meeting to discuss next month's pay.

His next stop was to meet the Special Forces Province Recon Unit (PRU) Advisor and his ARVIN PRU Chief and ask them to send a few PRU Tiger Scouts dressed in civilian clothing to watch the men working on the painting and clean up of the PIC buildings. The Province Interrogation Center (PIC) men were not organized and had no real leadership. There was no sense of urgency in their actions. In the past they had only used impact torture interrogation on prisoners, and had no concept of how to obtain information with out torture. Gilbert wanted to find those men who would follow his orders without question because when the next VC terrorist was captured for interrogation there was going to be some new leadership in command of the Province Interrogation Center (PIC). He had to get complete control fast as his interrogation techniques did not involve impact torture but complete cooperation from everyone in the Interrogation Center.

The rest of the day was spent interviewing a new combat interpreter that had been sent to assist him. All the interpreters were better educated than most Vietnamese and received more pay. The better educated they were the more arrogant they seemed to become. As Gilbert didn't use impact interrogation it was necessary that the combat interpreters must repeat exactly and correctly what the prisoner said and not what he thinks Gilbert wanted to hear. Most interpreters have a tendency to try and run the interrogation. Gilbert interviews prisoners as if they were applying for a job or wanting to join the local Province Recon Unit (PRU). It was the good guy versus the bad guy type of a police interview.

During WW-II Gentle Ben, the Army aviator assigned to support the PRU Tiger Scouts used three Japanese interpreters as he interviewed Devine Wind Pilots on a Japanese suicide base named Fujigaya Air Field, Chiba near Tokyo, Japan. He would have one of the interpreters in the room working and two out of hearing range with a guard. Every five minutes or so he would change interpreters and ask the last question again. Gilbert used the same technique only he used a tape recorder and studied the questions with the answers as he rested in the evening to be sure the interpreter was not putting some of his own ideas into the interrogation.

One time Gentle Ben was bragging to Gilbert on how he only had 40 soldiers with none over the rank of Army Corporal in charge of 1500 Japanese Soldiers as prisoners after WWII ended. He hired their Commander as his Japanese General Manger and paid all the soldiers to work for the US Army. Next he furnished a handful of rice per day for each man, and if he had a family two hands full plus a hard bread roll made out of wheat floor for each member of the soldier's family. Finely he had wooden clubs issued to his newly employed Japanese police as they turned in their military weapons. He established a dark to daylight curfew and used the Japanese court system for punishment. To keep everyone busy he modernized the base laundry, built American family housing, barracks for the US Army soldiers, farmed the vacant land for growing crops, and built a modern sewer system. Gentle Ben was so pleased with the control he had of the Japanese reparation funds he hired Japanese attorneys and purchased the land of the Golf Course that had been the Fujigaya Air Field.

Good things never last long in the political world, and someone from Tokyo, GHQ found out what he had done. In one twenty-four hour period the US Army Attorney General had more charges filed against Gentle Ben than he could carry in a barrack's bag. While under base arrest the Army lawyers were circling like hawk over a road kill.

Lucky for Gentle Ben a young AG Captain had just arrived from the United States who knew nothing about Japanese law or anything about Japan. Without asking he had all the deeds transcribed into English to see if he could find new charges to file. Low and behold Gentle Ben's name was not on the Deeds. All the land was deeded to the United States Army in behalf of the United States of America. Gilbert asked, "What happened to you?" "Nothing. Well, almost nothing. They replaced me with a US Army Colonel and about twenty officers plus 600 soldiers. I learned later that the AG Captain also found that the Japanese Army had leased the Golf Course. It wasn't Japanese Army property, and this made the sale plus land transfers valid. The US Army kept the land." Gentle Ben answered and then said, "My Japanese lawyers hadn't steered me wrong, and they sure earned their fee."

Then Gentle Ben asked Gilbert what he was going to do once the Army found out that that the Communists were being defeated and the areas pacified where ever he had taken control. "I'll bet you that when they realize what you have done the Army will take over." Gentle Ben said. "I'll just go home and do as you did. Nothing," Gilbert laughed as he answered.

Besides being better educated the interpreters received more money than a Captain in the ARVIN Army but they were usually draft dodgers or had political connections. Actually Gilbert picked and controlled them much the same way he did the PRU Tiger Scouts. If they agreed to do it his way he would pay them more, and then to clench his control he would build them a hutch within the guarded compound for their families. With his family secure and no rent it wasn't long before the interpreters wouldn't breathe in or out until they were sure Gilbert approved.

Gilbert and Monk stopped at the Province Interrogation Center (PIC) office the next day and just looked around without speaking to anyone before they left. After three days the buildings were painted white inside and out plus the grass was cut. Each day they would visit the PIC for about fifteen minutes but always at a different time of day. Gilbert would take a few of the personnel records with him, and trade them for others the next day. He did this until he had read all the files with out saying a word or making a comment. It was enough to put the fear of God in them. By the third day their arrogant attitude had disappeared.

Everyday the Vietnamese interpreters and interrogators were working harder and looking sharper. Now was the time for a little sight seeing so he rode around the town with his bodyguard and Combat Interpreter. It was actually a very serious business for Gilbert as he was trying to find a pattern to the Dark Forces night mortar attacks. By checking the damaged buildings and targets he was trying to estimate the locations of where the mortars were being fired from. It wasn't the mortars that interested him but where his PRU Tiger Scouts could set up ambushes to capture the Viet Cong who were firing them. He needed to obtain information on where the ammunition was stored and to identify individual VC. The transportation of a heavy mortar would take one or more men or at least a bicycle and he guessed that they wouldn't move it far from its hiding place.

As he drove Gilbert would keep looking down at a map held on the steering wheel. He would stop now and then to make to make a mark on the map. The interpreter asked what he was doing and Gilbert said, "I'm marking down restaurants where I may want to have dinner. I never eat at the same place twice in the same day. Actually I won't know the restaurant myself until I get hungry." His combat interpreter just nodded.

Monk could care less as he had his begging bowl and didn't eat in restaurants. He would put on his orange robe of a Buddhist monk, lay the M-16 on the sidewalk, and squat outside the door with his bowl waiting. Gilbert never said anything but just watched. If the restaurant owner didn't fill the bowl before his meal was served he would pay to have it filled with the best meat and rice. Gilbert realized in Vietnam it wasn't just having a bodyguard to protect you but having his respect. A full belly helped.

Gilbert had survived for almost two years now while beating the Communists at every turn. He was a prime target and on their list of most wanted to get rid of. Of the thirty-six contract intelligent officers in his CIA class there were only three left in Vietnam, and he was also an interrogator. Actually he was not hated by the enemy but only feared. They respected him and feared him at the same time. Kill him they surely would but it was more like trying to removing a road block on a highway just so they could move faster. Gilbert and the Viet Cong understood each other. Bill Colby enjoyed watching Gilbert work. He knew that it wouldn't stop the Paris Peace Talks or change the power struggle in the United States between the Dark

Force and the Bright Force but he just wanted it recorded that the Communists could be stopped.

Now that the Province Interrogation Center was ready for work, all that had to be done was to have the PRU Tiger Scouts capture a few Viet Cong (VC) to be interviewed. The VC that Gilbert really wanted to talk too were the same Communists who had been firing mortar rounds into the compound and the government buildings in the city. When he told his Special Forces Army Advisor of the PRU Tiger Scouts and the Province Interrogation Center (PIC) people they informed him that there was one small problem. It seems that the Korean Marines of the Whitehorse Division were cutting the heads off of all the Viet Cong captured. The Korean Marines were the main military force in the area, and all prisoners were turned over to them. They explained that this was why the PIC never had anyone to interrogate. Gilbert understood the problem because even he couldn't get information from a corpse.

Then Gilbert explained his plan on how to capture the Viet Cong when they were firing a mortar. He also told them where he thought that the mortars were being fired from. He said, "Have three squads of PRU Tiger Scouts set up an ambush at these points which he had marked on the map he had drawn. It may take one or two days. Or maybe even three but the next time they fire a mortar you should be close enough to capture the Viet Cong firing it. Once you have them do not turn them over to the Koreans or even tell them. It's hard to interrogate a prisoner with his head cut off. Isolate any prisoners you capture, and don't do any interrogations until I give you the word."

The next day Gilbert, Monk, and his combat interpreter drove a jeep to the Korean Whitehorse Division, and as the guard at the gates saluted, Gilbert asked him where the S-2 shop was. It was a log bunker built underground with dirt floors, log ceiling, and sandbagged walls. The first motor round hit as they walked in. They found themselves standing in a large dark room with dirt floors and low ceiling. The impact of the mortar rounds created a dust cloud that filled the air, and it was hard to see or breathe.

A Korean Captain, the S-2 officer, came out of the next room and upon entering he sneezed. Gilbert spoke without thinking. In German he said, "God Bless You." The Captain asked, "Do you speak German?" They stood chatting in German as Gilbert spoke no Korean and the Captain no Vietnamese. They never got around to speaking anything but German, and when they parted Gilbert never guessed that the Captain spoke and understood English. They finely got around to why Gilbert was visiting the Whitehorse Division.

The Captain asked how he could help. When Gilbert suggested that any Viet Cong they captured be sent to the Province Interrogation Center (PIC) before they were killed; the Captain just shrugged his shoulders and said that would have to be the Commander's decision. It hadn't been done before and seemed like a lot of trouble for nothing. Gilbert asked, "Can I meet with the Division Commander?" "He's busy right now but if you throw a party as a celebration for moving into the Division Area, and invite him as your Guest of Honor he may help," the Captain answered and then asked, "What do you do with the Viet Cong after you are finished interrogating them?"

"I enlist them as Province Recon Unit (PRU) Tiger Scouts and use them to capture more VC." Gilbert said. The Captain looked at him as if he was nuts. Then he said, "Better forget to tell the Colonel. We just kill them and anyone around them at the time. If they make a mistake and take out one of our trucks with a roadside bomb we burn the houses all around the site and kill everyone living in the area. Those mortar rounds that hit us as you came in was a big Viet Cong mistake. When you leave you will notice that there are no buildings or roadside stands within a thousand feet of the road. As we have been visiting our troops have killed everyone and are burning the buildings within the zone that the mortars could have been fired from."

All this had been spoken in German. Gilbert's combat interpreter and bodyguard were both amazed as they drove out the gate to see everything leveled. There were a few bodies lying around and it looked like they were going to be there until their Vietnamese families came to get them. Gilbert told Monk and his interpreter what had happened. Then added, "We'll not stop until we are in the city. I don't think anyone will be around until the Koreans are back inside their compound." The Whitehorse Division S-2 was friendly and acted like he would help. The only problem was he couldn't understand the need or why Gilbert wanted live Viet Cong.

When Gilbert had first moved to Tuy Hoa he was assigned a room in an old French Mansion facing the South China Sea. Unfortunately, because of his popularity amongst the

Viet Cong, and being the prime target for mortars no one wanted to live in the same house that he did. The house was fully furnished except for his personal items and that included an old Chinese woman cook who came with the house. She lived in a room near the kitchen, purchased food, household goods, and prepared his meals. She also supervised the Vietnamese maids, wives of the PRU Tiger Scouts, who came in during the day to clean. Before when he had to eat in restaurants all the time it was necessary to be on constant alert for a sapper (terrorist) attack.

The concertina wire of the PRU Tiger Scouts' Compound included the house and grounds to Beach Road. While this did not stop the mortar attacks or rockets it made life a little easier as there was only one gate that faced the road across from a long white sandy beach of the South China Sea.

Gilbert still needed to identify the Viet Cong leaders and find out where they were going to strike next. He needed to get his Province Interrogation Center (PIC) operational, and that meant capturing Viet Cong. His advisor of the PRU Tiger Scout's came to his house and met him just after he finished breakfast one day. He reported that they had captured one Viet Cong but he had been wounded, and it would be at least a week before he could be turned over to the PIC. The prisoner was captured as he tried to set up a mortar in one of the locations indicated on the map. Unfortunately at the other two locations all of the Viet Cong had been killed in the ambushes. The good news was that his plan had worked, and they had captured three mortars with twelve rounds of ammunition.

The Koreans were still his best bet yet to get prisoners for interrogation. Gilbert decided that he should drive back out to the Korean Base and extend his personal invitation to the Division Commander for a party at his house as the guest of honor on the following Saturday night.

Gilbert, his bodyguard, and interpreter were driving on the road towards the Korean Base, and just at the edge of town with small shops on both sides of the road his jeep was ambushed by the Viet Cong. Gilbert went off the road and ended up with two vertebrae broken in his back, and a trip to the Tuy Hoa Air Force Base hospital emergency room. The combat injuries reports were censored by someone in the Embassy and the medical files were forwarded to Nha Trang Region II Hq of the American Embassy. After the Air Force doctors taped him up they told him that he should stay in bed and rest for a week or two.

Even with the pain this was an impossible condition for Gilbert to be in. There was no way that he could take the time off to stay in a bed. Besides, he had to prepare for the party to honor the Commander of the Whitehorse Marine Division. First he had an invitation delivered to the Commander of the Whitehorse Marine Division from the MACV Headquarters through the Korean Embassy for a personal dinner for him and his staff at Gilbert's house. The party was designed so that the Colonel was the highest ranking guest present but with just a hint of Gilbert's power in the background. He needed the Koreans on his team but not where they could run with the ball, and the invitation was part of the plan.

He knew from going on missions with his Special Forces Advisor and Chief of the PRU Tiger Scouts in Boston Whalers on the Mekong River how to perform reconnaissance with M-60 machine gun fire. He had also read reports of how Gentle Ben always had his gunners' recon by fire when landing a helicopter in a free fire zone. There would be no problem with security or an attack from the Viet Cong during the party as the Koreans had their own system of reconnaissance. They had trucks with quad 50 Cal. machine guns mounted on them so they could fire in a 360 degree circle. Each truck would lead the way followed by a marine personal carrier, then the Commanders jeep, and more personnel carriers. The driver of the lead truck only knew two speeds. Full throttle and stop. The men manning the Quad 50 Cal machine guns shot at anything any one could hide behind, rocks, bushes, or anything they thought looked out of place. It has been said that the Viet Cong would run out onto the road and dig up mines that had been placed if there were a Whitehorse Marine Division truck on the road.

It would be six days until the party, and Gilbert only had one Viet Cong waiting to recover enough to be interrogated. It wasn't enough besides he wanted to learn more about how the Dark Forces controlled the city at night. Gilbert always drove his jeep as if it had only two speeds; stop and fast as hell. Everyone was working getting ready for the big party. His Chinese Cook and head housekeeper was too busy to even think about cooking for him until after the big party. The PRU Tigers Scouts were patrolling the area around the compound, and the PIC personnel were making ready as if it was President Nixon and Bob Hope coming to dinner.

The Korean S-2 had already sent notice that the entire Division would be on alert and have patrols in and around the area. Even the Communist VC must have decided that this was not the time or place to start anything.

With nothing to do until after the party Gilbert drove his bodyguard and his interpreter around town to see if there was anything he could see that was not just as it should be. Gilbert noticed a young woman standing outside of an aid station. He slammed on the jeep's breaks, skidded, and stopped. She was holding her arm and crying. She was in great pain. He got out of the jeep and asked her if he could look at her arm. It didn't seem to be bleeding. The arm was broken and at an angle that if left alone would never be used again. Gilbert finely got her to tell him that she had been questioned by some men who had grabbed her a few nights ago. She said they had twisted her arm until it broke before asking her to get information from USAF Officers at the Tuy Hoa Airbase.

A CIA Interrogator is not just someone who works in dark rooms, and asks questions but is a person who uses the tools at hand to obtain information. Gilbert is one of those who know when the Bright Force has given him a tool. He asked if she would like him to have an American Doctor fix her arm in an American Hospital. She was a very young pretty girl. It was no wander the Viet Cong thought she could get information from American Officers. She looked like a living doll. "What do I have to do?" she asked. "Just point out the men who did this?" Gilbert answered. He then told her that he would feed and pay her for each VC she pointed out plus a smaller amount for each cache of weapons they found. As she still hesitated Gilbert said, "I'll even give you a new ID card and fly you to Saigon or any place you want to live in Vietnam after you identify all the Viet Cong who hurt you."

If there is a secret to Gilbert's success it's that he never makes a promise he can't keep, and always keeps his promise. Gilbert had been ambushed the week before and the Air Force doctors who worked upon his back had told him to come to them anytime he needed their help. Even though it was supposed to be a secret the American doctors and nurses all had agreed to help at any time. No one in their right mind would have guessed that this is what they were thinking when they offered to help him. The CIA Agency's old Dr. Slocum gave Gilbert his famous pain pills made from cobra venom called "Cobrison". They really worked, and he was back in the fight.

When he showed up at the USAF Base Hospital with a beautiful Vietnamese doll, and met with the doctor from Ohio, he just asked, "Is this another secret operation. How can I help?" Gilbert answered, "It's a secret alright, but it's even more than that." Then Gilbert told him who he was and what he did. Then he said, "This is a Viet Cong girl who has agreed to help me capture the people who did this to her. If you can help her it will save a lot of Americans lives." The Captain called in three other doctors and Gilbert once again told everyone who he was also what was needed. One doctor said, "We can't have her in the hospital as the Base Commander would have a Tizzy." Another doctor said, "There is a storage trailer out back where we can keep her after the operation. You can bring her with you for your check ups so we can see her. You can do this can't you?" He asked.

This was one time being a CIA National Police Adviser really helped. Gilbert's pass or ID (Walk on Water Card) gave him permission to go anywhere at any time, and have someone with him with no questions asked. The doctors couldn't perform the operation during the time the Hospital Commander was present but somehow he must have found out for he took off to visit friends for a few hours. Long enough for the doctors to operate and repair the damaged arm. Then the nurses and ward men took over. There was always someone going out to the supply trailer where they had hidden her. The doctor from Ohio told Gilbert that she had come through the operation just fine and could be moved in three days. Only the nurses gave him trouble. They kept coming up with reasons why he couldn't move her. They bathed her, combed her hair, and fussed over her. Gilbert was there when she woke up in the storage room. She just held his hand and cried. It was tears of joy not pain.

The three days were up, and Gilbert had to move her to the Province Interrogation Center (PIC) because she was part of his plan to influence one Korean Division Commander. It wasn't the doctors but the nurses who were objecting. Even with the promises that he would bring her back for check ups, he thought that the Korean Marines might have to help him get her away from the hospital. Gilbert had to get her back under his control before one of the doctors proposed or he would have had wedding bells instead of an informer. The Province Interrogation Center (PIC) interrogators had painted a room and found an Army cot for her to sleep on. They put reed rugs on the floor and had even found a box for a dresser. Someone even donated an old used Army foot locker and a chair.

Gilbert picked out a young good looking policeman to go on the search operations in the city with her. Just like the special police operations in Ben Tre City they had hand signals for the girl to indicate who the Viet Cong were. By the day of the Colonel's party she had identified more VC that were captured by the PRU Tiger Scouts than the PIC Interrogators had ever seen or interrogated before.

The day of the big party came and around 1600 hours with a roar of engines the Korean Marines arrived. Their Commander and his staff were met at the gate of the compound by Gilbert with his bodyguard while the road from the gate to his house was lined with PRU Tiger Scouts shoulder to shoulder all the way to the front door. Gilbert greeted the Colonel by speaking in German to his Captain the S-2 (Intelligence Officer) who acted as an interpreter. As they passed each platoon of PRU Tiger Scouts they would brake formation and form into six man squads. With a company of Korean Marines around the house and over two companies of PRU Tiger Scouts patrolling the compound the dinner begin.

The Chinese cook outdid herself and prepared a seven course meal. The division commander, sat at the head of the table with his S-2 next to him. Gilbert was talking in German to the captain, and he would translate into Korean for the Colonel. As the dinner proceeded the Colonel noticed a Jim Bowie knife that was mounted on the dinning room wall for decoration. The Captain asked if the Commander could hold it. Gilbert took the knife from the wall and presented it to the Colonel while holding it in both hands palms up. The Colonel took the knife and thrust it in all directions, making as if he was using it as a stabbing weapon. He returns the knife to Gilbert in the same manner as it had been presented to him. Gilbert refused the knife telling the Colonel that this was really a gift of friendship. The Commander was grinning from ear to ear. He spoke to his Captain (S-2) who then turned and said, "The Colonel would like to see this PIC that wants live Viet Cong."

The dinner was officially over. So Gilbert led the Colonel out to his Jeep and with his PRU Tiger Scouts as guards took him out of the compound to the National Police Compound where the Province Interrogation Center (PIC) building was located followed by all the Korean Marines. There were almost four dozen marines following them. In Gilbert's Journal he wrote that it was as if General de Gaulle had left the building in Paris. The guards were 12 blocks deep in all directions. The only difference was that Gilbert was driving at full speed (flat out), and only he knew their destination. All the Vietnamese were running trying to get out of the way. It was pure chaos.

The Province Interrogation Center (PIC) was a white "U" shaped building. All the Interrogators were standing outside of their interrogation rooms and as the Colonel walked in he inspected them never speaking but just nodded at each man. In the last room the VC informer was standing with her arm in a sling. She was sure a pretty thing standing there with a big smile on her face. She bowed from the waist, and he bowed back. Gilbert turned to the Captain and said, "Tell the Colonel that she has captured 13 VC in the last two days, and I expect that she will do much better when her arm heals."

The Colonel turned and spoke in perfect English, "You will receive all of the Division VC prisoners that we capture from now on. Do you want the dead ones too?" Gilbert turned to the S-2 and said in German, "He speaks English". The Captain laughed and answered in English, "All of his staff officers speak and understand English."

The next day Gilbert phoned Bill Colby in Saigon and reported everything that had happened. His boss was so excited that he kept interrupting asking how many VC were captured. "Thirteen," Gilbert answered. The Koreans S-2 said that they will send over three more VC tomorrow but they are in bad shape. "What are you going to do with all the prisoners?" Colby asked. "All those I can't turn louse I'll return to the Koreans," Gilbert answered. And then added, "I'll start teaching the PIC Interrogators how to get the information we need. The Communists don't have a lot of choices. Become a prisoner of the Koreans and have their head cut off, or after we post their names as members of the local PRU Tiger Scouts being skinned alive by the Communists or really joining and getting paid to capture other VC." "Take Care. The Viet Cong really don't like you, "Colby said as he signed off.

A few days later the S-2 for the Korean Marine Division phoned, and told Gilbert that they had a VC prisoner for the PIC. Gilbert and Monk jumped into his jeep and headed out of town toward the Korean Base. Gilbert drove his usual speed of flat out and was getting 70 miles per hour out of the jeep. Suddenly he could see an MP Jeep in the road ahead and someone waving for help. He slammed on the breaks at the same time telling Monk his bodyguard to prepare for action. As they skidded to a stop Monk jumped out swinging his automatic M-16 with the long clip looking for a target.

As they skidded to a stop an MP Captain walked up and said, "We clocked you at 70 mile per hour with our new radar gun." He didn't have a chance to say anything else before Gilbert asked, "Are you under attack?" Upon the Captain asking him again if he knew how fast he was going Gilbert hollered to Monk saying, "Get in. If this MP tries to stop us shoot him." With the tires spinning they took off a full throttle.

When two Korean Marines threw the prisoner into the back of the jeep they had his hands wired together behind his back and then to his feet. The man looked more dead than alive as they prepared to leave for the PIC. Gilbert thanked the Korean Captain and by the time they passed through the gate the jeep was doing 50 miles per hour. They had been ambushed once before on this road and knew that the VC would never mine a Korean road. An ambush they would attempt but only against American or Civilians. The cost was too high to attempt to blow up a Korean vehicle. Speed was Gilbert's best defense against an ambush. Get through the kill zone before they could identify him or his vehicle.

As they drove down the road Gilbert told Monk, "Take our canteen and give him a drink of water. After he drinks pour the rest on his head. See if you can untie his legs and sit him up." The prisoner didn't know it but Gilbert was already starting his interrogation. He always played the tough good guy and it worked.

The next day an MP Colonel and his Sergeant drove into the National Police Compound looking for the Police to help find an unmarked jeep. He didn't know the drivers name but only that the jeep was an unmarked vehicle with no license or marking of any kind. As they looked for a place to park the Sergeant spotted Gilbert's jeep. The Colonel was driving, and he parked next to the unmarked jeep saying, "Sergeant, stay here and stop anyone trying to leave with that jeep before I get back." As they drove into the Police Compound they had passed a "U" shaped building painted white being guarded by a few PRU Tiger Scouts posted around the building dressed in their black pajama uniforms. There were also a few white mice (police) in dark pants, white shirt, and uniform caps walking or moving around the buildings. Every thing looked peaceful much like any police headquarters on a hot summer day.

As the Colonel walked into the office door next to the unmarked jeep there suddenly appeared two squads of black uniformed men carrying automatic weapons who took up positions between the police building and the parked Jeeps. A US Army Sergeant dressed in a Special Forces combat uniform walked over to the MP Sergeant and said, "Don't get out of the jeep. Slide over or standup and get into the driver's seat. Move your jeep into the roadway facing toward the gate. Don't step out of the vehicle and keep the engine running. When your Colonel gets in drive slowly out of the gate. Don't turn or stop. I'll notify the guards that you won't be stopping."

The MP Sergeant just stood up and stepped over into the driver's seat. He backed the jeep up until it could be turned with the passenger's side facing the door where his Colonel had entered into the building. One of the men in the dark pajama uniform laid down and moved under the unidentified jeep while another raised the hood. It was apparent that they were searching the vehicle looking for booby traps. All the others had moved placing the MP jeep in their kill zone. The Special Forces Sergeant had disappeared. The MP Sergeant noticed that there were no longer any uniformed police officers in sight or anyone other than black uniformed men carrying automatic weapons, and their guns all seemed to be pointing directly at him.

There was no sign on or near the office door indicating what business was conducted therein. As the Colonel walked in he saw a long office with a movable portion blocking the view of the back part of the room. On the right side was a bench with a black uniformed Vietnamese sitting with an M-16 laying on the bench beside him. On the other side of the room in front of the portion was a large desk with a Vietnamese secretary working on some papers. On the wall were two charts and a map of the province. One chart was labeled in English "PRU Training Schedule" and the other was labeled "KIA Flag/Date." In the map were colored pins at various locations but no writing of any kind. All three items were constantly on Gilbert's mind. Training of the PRU, and his duty to put a flag with a medal on the coffin of those killed. The map was the most important as it tracked the movements of the known Communist Viet Cong that still hadn't been captured.

The Colonel saw a pretty woman near thirty wearing a gold chain with a small gold cross around her neck dressed in traditional silk trousers covered by a bright pink split skirt and a flowered blouse setting at the desk. She stood up and greeted him asking, "May I be of assistance Colonel?" He was already mad as she greeted him in perfect English even before he noticed that the soldier setting on the bench now had the M-16 on his lap pointing at him. "I want to see your Boss," he said. He was an MP who had been around a long time, and knew when he was outgunned by someone who wasn't afraid to use his weapon. He didn't know that this was Gilbert's combat interpreter who was just waiting while Gilbert finished reviewing the audio tapes made at their last interrogation before they went to the PIC to interview another prisoner. He could hear voices coming from behind the portion like an interview of how a Viet Cong attack was going to be made against the Whitehorse Marine Division Firebase. His face started to get red as he started forward. The man with the M-16 had moved across the room so the secretary was out of the line of fire. The Colonel froze as his mother hadn't raised any stupid children. He had to see who these plotters were. He didn't know it was Gilbert playing the tape recording of a prisoner being interrogated. All the Colonel knew right then was that the secretary was smiling as if candy wouldn't melt in her mouth.

She didn't walk into the killing zone that the Combat Interpreter had established, but was still able to slide into Gilbert's office which was behind the portion and announce that there was an MP Colonel who wanted to see him. When the MP Colonel came into view he was a giant of a man and appeared to be strong enough to break Gilbert in two parts. The Colonel was looking down at a man dressed in the same black uniform pajamas with no rank as the man in the front of the office, and the soldiers in the police yard where they had parked. Just to the left sitting against the wall was another man dressed in the same type uniform but the M-16 was slung over his right shoulder with the barrel pointing towards the floor. There was no one else in the room except a tape player still running with the voices of the people he had heard talking. Behind the desk was a window making him look into bright sunlight light that made it hard to distinguish the features of the man at the desk. This was a good thing as he hadn't become an MP Colonel by being dumb. It gave him time to realize that he was not only out gunned but in deep shit.

Gilbert didn't stand up but remained seated and asked, "How may I help you?" "Yesterday you threatened to kill one of my officers." He said and then added, "Who the hell do you think you are?" "That is perfectly true. You do not need to know who I am. Whoever ordered a speed trap to be placed in a combat zone is plain stupid." The Colonel acted like he was going to answer but Gilbert said, "One more word and I will have you thrown out of this province. Now get out of my office." When the Colonel spoke and Gilbert answered Monk knew from the tone of his boss's voice that he wasn't happy. He came up out of the chair with his 10" switch blade knife open, and the point under the Colonel's chin. The Colonel was a big man but Monk could look him straight in the eye and was ready to kill. The Colonel knew he was in trouble but had no idea that he wasn't dead only because Gilbert still had his hand open on the desk.

As he turned and went out the door he couldn't help noticing that the secretary was busy working on some papers and didn't even look up. The black suited soldier was now standing next to the charts with his M-16 pointing directly at him, and his finger was on the trigger. He also noticed that the man could fire without fear of hitting anyone else in the office. The Colonel was still mad but not where he didn't realize that he was lucky to be able to walk out of this office.

Stepping out into the bright sunlight he noticed his MP jeep parked in front with its engine running and his Sergeant in the driver's seat. Climbing in he said, "Let's get the hell out of here." The jeep started moving just above a walk toward the gate. "Move it," the Colonel said. "Sorry Sir but I've been instructed that if I stop, turn, or go over 15 miles per hour it will be signal for them to open fire." Then the Colonel noticed that the black uniformed soldiers were in positions where the jeep was in a kill zone all the way to the gate, and there were no police uniforms in the area. There were six police officers in the gate house and at least that many in the guard tower covering the gate. They just waved them through smiling and laughing.

CHAPTER FOUR

As they drove away from the National Police Compound the Sergeant asked, "Did you find out who was driving the unmarked jeep?" "I think we met but I'm not sure. I'll say one thing. I'm sure that our Captain wasn't threatened. That man only makes promises. The Captain is lucky he wasn't killed." As they drove along the Colonel kept feeling his neck. "Can you see any blood?" He asked. "I don't see anything. It looks normal to me." The Sergeant answered. "I think we are very lucky to be driving back to Tuy Hoa. I have a feeling that we could be in body bags and no one would even know what happened. I'll have to ask the military police command in Saigon who these soldiers are, and then I'll have another talk with the man who drives 70 miles per hour in a jeep." the Colonel said. "Sir, if you go back I'd sure like to be someplace else. It's not a fun place to visit," the Sergeant said. "If I go back I'll go alone or maybe after a B-52 bombing mission with that police compound as the target." The Colonel answered.

After the MP Colonel had gone out of the office Gilbert shut off the tape recorder, reset it, and walked out followed by his bodyguard, and combat interpreter. As they walked across the compound toward the Province Interrogation Center (PIC) building Gilbert realized that the MP Colonel must speak Vietnamese for he had looked at the recorder with understanding of what was on it. He looked at his watch and asked, "What time did they start the interrogation?" "0900 hours as you directed," the Interpreter answered. "That dam fool MP Colonel has delayed my stopping the impact interrogation. That's the second time he's slowed me down. Remember just repeat what I say and what the prisoner answers. Don't add a word or speak until I start the tape recorder," Gilbert said.

As they entered the building they saw that the Viet Cong prisoner was naked, and strapped on a wooden table. Wires were attached to the VC's testicles leading to a hand cranked field phone. The table was tilted so the man's feet were above his head, and it appeared that two interrogators had been beating the soles of his feet with thin bamboo strips. The man's feet were bright red, and blood was dripping onto the floor. It seemed that Gilbert's plan to prevent torture needed a little more work. When he hadn't arrived on time the Vietnamese Interrogators had started their impact interrogation. It was the same procedure as the North Vietnamese Communists interrogators used. When the interrogators got tired of beating the prisoner then and only then did they start asking questions.

The first thing that Gilbert noticed was that there was no tape recorder or anyone taking notes. He was mad and wasn't play acting when he ordered the interrogation stopped. He cursed and while not speaking to anyone direct said, "Stupid asses! Fools!" and other unkind words in three languages He ordered someone to bring a bucket of water for the man's feet and to remove the wires. Then he suddenly noticed that the combat interpreter was repeating word for word everything he said in Vietnamese. The Province Interrogation Center (PIC) interrogators all acted like they wished they were someplace else.

Now Gilbert was not just a CIA Interrogator but a man who never forgot that the Communists were the real enemy and while he couldn't stop their lies about Americans he could stop them from hurting and killing the people in Tuy Hoa. As Gilbert raged he walked about the room, and everyone he looked at Monk would point his M-16 while watching his hands waiting for the signal to fire. Even when he looked at his interpreter Monk was ready to kill. This was one man who didn't even look like he wanted to do anything but repeat Gilbert's words.

The prisoner had been taken off the interrogation table, and was sitting in a chair with his feet in a bucket of water listening as Gilbert gave everyone hell. He didn't speak but just sat watching as one of the interrogators tried to clean up the blood. Without looking at him or even speaking to the Viet Cong Gilbert turned to the head PIC Interrogator and said, "Phone the Korean S-2. Tell him to come and pick up the prisoner. I can't use him. His feet are injured so I can't use him as a Province Recon Unit (PRU) Tiger Scout. He doesn't know anything so send him to the Koreans." The PIC Chief said, "You haven't asked him any questions. How do you know he doesn't know anything?" Gilbert answered, "If he knew anything he would have told you during your interrogation. If he needs water give it to him and let him eat. If he wants to speak to me and can walk to my office I'll be there until the Koreans pick him up." Then he turned and left the PIC building followed by his combat interpreter and Monk. Everyone looked stunned as they left the building except Monk, and he was grinning.

As they entered the police office building he told his secretary to bring in another chair as they would be getting a guest soon. Then he put the tape recorder on the desk and sat down just as his secretary announced that someone with rags on his feet would like to speak to him. Gilbert didn't move as the prisoner walked in, came to attention, and saluted. Gilbert reached over and turned on the recorder before returning the salute. "I would like to join the PRU Tiger Scouts," the Viet Cong said. "It will take weeks before you can work as a soldier why should I wait?" Gilbert asked.

"The North Vietnamese Province Sapper (terrorist) Battalion has been reinforced, and they are going to attack Tuy Hoa Air Force Base in four days. They have enough sappers who have volunteered as suicide bombers to destroy most of the USAF fighter bombers on the Airbase," he answered. "Do you know the location of the attack point?" Gilbert asked. The prisoner nodded and said, "I can even tell you the time when the attack is planned to take place." Without answering Gilbert called to his secretary and told her to notify the Special Forces Sergeant Advisor of the PRU Tiger Scouts that he had another recruit. Also that the new recruit should be placed on light duty until he could wear shoes, and to put him on the payroll as a PRU Tiger Scout, starting today. He stood up and as he started out of the office his Combat Interpreter and Monk followed him.

Gilbert looked at his watch as they left the building and it showed 1100 hours. When they got in the jeep he told his combat interpreter and bodyguard that he would be having lunch at the Tuy Hoa Airbase officers club, and afterwards he would be visiting the hospital for a checkup on the back injuries he received when he was ambushed by the Viet Cong. He got in the driver's seat of the jeep and drove over to the Province Interrogation Center (PIC) building. After stopping he told his interpreter to fetch the Viet Cong female informer as the doctors would want to check her arm.

Besides he was going to invite everyone in the hospital to his house for a party, and once again she was the bait so to speak. She was the young girl that everyone at the Air Force Base Hospital called the Vietnamese Doll. During lunch Gilbert told her that on Saturday there was going to be a party, and she was the guest of honor. He told her that this coming Saturday she would also be released from the Province Interrogation Center (PIC) and on Sunday an Air American aircraft would fly her to Saigon. Also he reminded her about his promise that she could live with her aunt in Saigon after they had captured all the Viet Cong who had hurt her. This had been accomplished, and all the doctors, ward men, and nurses who had helped her would be invited to the party. She looked very serious and didn't smile. Finely she asked, "Do I have to sleep with all of them?" Gilbert was drinking from his glass of water when she asked the question. He started choking and sputtering water as he said, "No, this is a thank you party." "I have no money to pay them. Won't they be unhappy if I don't give them something?" She asked. Gilbert was at a lost for words. He hadn't thought about how the Vietnamese expected a gift when honors were given. Then he said, "I'll have the girls who work at the Tuy Hoa Hotel come to the party to keep the men happy. I'll pay them and the hotel." She just nodded and went back to eating lunch acting as if this solved everything.

His Interpreter had taken Monk's bowl and had it filled plus a plate of food for himself to eat on the beach in front of the officers' club. After lunch they went to the Tuy Hoa Hospital where he and the VC girl informer got out of the jeep as his interpreter slid into the drivers seat. "Pick us up around 1500 hours," Gilbert told him.

As they walked into the hospital everyone greeted them with smiles while directing them to the x-ray area. As the technician was taking pictures of the Viet Cong girl's arm Gilbert asked him where he was from. "I'm from Colorado Springs," he said. "What kind of work does your father do?" Gilbert asked. "He makes race forms for greyhound racing dogs," the technician answered. Then Gilbert told him, "your dad is an Italian and your mom is a short little lady. She is a wonderful cook. I've visited your home in Colorado Springs." Gilbert went on to describe his family's home on a shady street. "When did you visit them?" the airman asked. Gilbert answered, "I always remember a good cook. It has been about twenty years ago." Then he asked, "Our Vietnamese Doll will be leaving for Saigon Sunday, and there is going to be a party in her honor will you come?" "I'll try but all military personnel are restricted to the airbase." the technician answered.

When Gilbert spoke to the doctors they also told him that no one was allowed off the airbase except with a special pass and never at night. As Gilbert was talking to the doctors he noticed that the nurses had gathered around the Vietnamese Doll, and one of the Vietnamese nurse's aides was acting as an interpreter. As they were all laughing he didn't do more than glance toward them. Then he asked,

"Can you go to a beach party in front of the officers' club this Saturday." They agreed that this would be the best place to have a party. One of the doctors asked who would be invited. Gilbert told them that all the hospital personnel were invited, and there would be a power speedboat to give rides for anyone wanting to go for a boat ride on the South China Sea. One of the nurses came over and asked if they could go on the boat if they wanted too. Gilbert answered yes but added that they may not have as much fun as the men. This didn't seem to be a problem so he just shrugged his shoulders and forgot about it.

When they were picked up at the hospital he told his Interpreter to drive back to the Officers' Club and went in to make arrangements to pay for the food and drink for the beach party. At the same time he told the club manager that he wanted a large bonfire on the beach, and that the party would be from noon until 2400 hours.

A few days later as Gilbert was sitting at his desk studying a map of the airbase his secretary announced that the MP Colonel who had been there before would like speak to him. He was busy trying to find a place where a boat could dock at the Officers' Club beach so his guests could get rides to the Province Recon Unit (PRU) compound and back. Monk had stood up and moved to where he could be seen by the MP Colonel as Gilbert stopped working and walked out to the front of the office.

"I'm sorry to have bothered you the other day. I would like to apologize. I've instructed my MP's not to stop your jeep in the future. If I can assist you in anyway please let me know," the Colonel said. Gilbert held out his hand

and as they shook, said, "Thank you for coming back. I'll remember your offer." The Colonel saluted and turned to leave when Gilbert said, "No more speed traps in my area." "I've already locked the radar gun in my safe. The backfire could get someone killed," the Colonel answered, laughing.

Gilbert went back to studying his map of the airbase. The boat ramp actually entered the Song Ba River's mouth where it joined the South China Sea, and after it was in the water a boat could be beached on the white sand in front of the officers' club and in front of the entrance to his PRU compound. The round trip would take under thirty minutes, and he could have one of his Province Recon Unit (PRU) Tiger Scouts run it back and forth all night long if it was needed. The boat would cruise far enough off the beach to stay out of the range of any small arms fire. A bonfire at the officers' club and another in front of the PRU compound would act like a beacon to guide the skipper of the boat.

Now all that was needed was to have his ARVN PRU Chief contact the hotel in town and make arrangements for at least five girls to come and take care of the men who needed that kind of attention. He instructed that at lest four tents be set up inside the compound gate along both sides of the driveway leading to the house. They were actually erected with two on each side of the driveway and far enough from the road to avoid any noise.

When the day of the party came Gilbert had his Interpreter and Monk drive the jeep to the Province Interrogation Center (PIC) to pick up the Vietnamese Doll. This was her last day of confinement and the party was in her honor. They brought her to the French Mansion where she would greet the guests as they arrived. Gilbert and a squad of PRU Tiger Scouts used the PRU pickup truck and boat trailer to move the speedboat to the ramp on the Song Ba River. It took an hour getting it into the water and ready for its first run to the USAF Officers' Club. There would be two men on the boat armed with M-16s who would take turns driving the boat. The PRU Tiger Scouts placed roadblocks at the bends of Beach Road, and kept three mounted patrols driving along the road so no one could interfere with the speedboat bringing guests to the PRU Tiger Scouts Compound.

On its first run Gilbert rode over to the Officers' Club where he could meet the guests as they arrived. After the boat returned with the first airmen to visit the PRU Compound everyone soon knew that the boat ride included a stop at a French Mansion for food, drinks, plus entertainment. As the sun started to fade the bonfires were lit, and now the boat was fully loaded on each trip. One of the USAF nurses came up and asked if they could go on the next boat ride. Gilbert was so busy he didn't even think it strange when four nurses escorted by doctors got in for a boat ride on the South China Sea. After they returned he suddenly noticed that every boat leaving had laughing nurses and airmen headed to his compound. One of the Vietnamese nurse's aides stopped and thanked him for the party bowing in the Buddhist manor. She asked, "Will this help satisfy the debt of the Vietnamese Doll?" All this time Gilbert had thought that it was a man's party and never realized that the Air Force Nurses would band together to help a pretty young girl pay a debt. It had really turned into everyone's party.

As the sun started setting a USAF Colonel walked up to where Gilbert was sitting and introduced himself as the Hospital Commander. Gilbert asked him to visit and have a drink. As they drank scotch on the rocks the Colonel said, "I've been watching your speedboat give rides all day, and for the life of me I can't see how they can have that much fun crowed together like that. You couldn't pay me to go for a boat ride." "They do seem to be enjoying the ride," Gilbert answered taking another swallow of scotch.

The bonfire in front of the Officers' Club had finely died down to just red coals, and the boat was unloading its passengers when the skipper called for him to join them and said, "That's everyone." After Gilbert boarded they went directly to the boat ramp on Beach Road, and as the boat was put on its trailer he climbed into one of the patrol jeeps. He was driven back to the house, and as Gilbert got out of the jeep he was met by his ARVN PRU Chief who announced that the Viet Cong Informer had moved out of the Province Interrogation Center (PIC).

"Where did she move to?" Gilbert asked. The PRU Chief just shrugged his shoulders and added that she had taken all of her things. "Does the Province Interrogation Center (PIC) Chief know where she is?" Gilbert asked. Again all he got was the shrugged shoulders. Gilbert was wide awake now as he realized that the man wasn't telling him everything he knew. As he stood in the entrance way to his house he begins to apply the methods of questioning that he used with the Viet Cong. Nothing seemed to work.

"Is Monk or my Interpreter here?" Gilbert asked. When he had left for the airbase Gilbert had told Monk and his

Interpreter to stay with the Vietnamese Doll. The man said that Monk had left for prayers and that his Interpreter was home with his wife. He told the PRU Chief to send a guard to bring his Interpreter to the house. Then he sat down in a chair and let the man stand as they waited. When his Interpreter entered Gilbert arose from the chair and asked, "What part of my instructions about staying with our Vietnamese Doll didn't you understand?" Again he got shrugged shoulders for an answer. "Is she alright?" This time the Interpreter answered saying, "I asked my wife and she said it was OK." "What does your wife know about it?" Gilbert asked, "She's a Vietnamese woman," was the answer. Now Gilbert had to stop and think. Here were two loyal Vietnamese men standing here looking as if they were cats who had just caught the canary and were waiting to be congratulated. He was mystified and really didn't know what to say.

Now it was his turn to shrug his shoulders and say, "Good night." Even if Gilbert didn't know what had happened he knew that Monk wouldn't leave the Vietnamese Doll if she was in any danger. He didn't blame her for moving out of the Interrogation Center (PIC) as soon as she could. Gilbert walked down the hall to use the shower to get the salt spray off from his day on the beach, and then walked naked down the hall to his bedroom. As soon as the bedroom door closed he felt that someone was in the room. He reached for a flashlight that was on the dressing table and turned it on. Sweeping the room with its powerful beam of light he looked to see if anything was out of place. As it swept over his bed he noticed that the mosquito netting was down, and someone was in the bed. Next he noticed boxes belonging to the Vietnamese Doll were stacked next to the bedroom wall. What was a man to do? He lifted the mosquito netting, slid under it, and got into his bed.

The next morning as his Chinese Cook served breakfast she was all smiles and asked if Missy was still going to Saigon. Monk had arrived, and following him was Gilbert's Interpreter. They were both smiling and seemed pleased with themselves. He sent Monk up to get the Vietnamese Doll's boxes, and told him to put them in the jeep. His Interpreter said, "All Vietnamese women pay their debts. His wife said it was better to pay one man than many." Gilbert just ignored everyone. Then he said, "After Air America leaves we will be going to Tuy Hoa Airbase to meet with the S-2. While I was at the hospital's party he stopped and asked me to visit him as they have a problem that we may be able to help with."

Their newest PRU Tiger Scout had told Gilbert that the airbase was going to be hit with mortar rounds and sappers (terrorists) were going to infiltrate the flight line. The man had identified another Viet Cong, and when this man was captured he confirmed that the attack was scheduled to start the following night around 0200 hours by mortars firing from the railroad track west of the airbase. In the excitement of multiple mortar rounds hitting in the barracks area the sappers (terrorist) would swim in from the sea to the beach at the east end of the runway. It was a one-way mission for the sappers (terrorist). After setting the explosive charges they were to attack the aircraft guards until the planes exploded. This was to allow the Viet Cong who were firing the mortars to escape with the weapons. Gilbert was going to visit the S-2 and inform him about the attack, and then ask about the problem that needed their help.

After crossing the Song Ba River the next turn off was to the left along the airbase's north boundary. It was about a half mile drive before the road turned into the main gate of Tuy Hoa Airbase. As they drove into the airbase the Air Police stopped his jeep and after checking his identification, used a hand held radio to notify the airmen in the guard towers that he had been cleared. It was a good security system, and when they arrived at the headquarters building Gilbert told his interpreter and Monk to remain with the jeep as the meeting would only be for a few minutes. He even told them that they would be going to the Korean Compound next and would eat lunch there.

Gilbert was by now a friend of the USAF S-2 and everyone greeted him with smiles as a sergeant escorted him directly into his office. There was a small meeting going on but everyone got up and left as the Lt/Col S-2 Officer greeted him. Gilbert told him about the coming attack on the airbase and said, "I'm on my way to the Korean Whitehorse Marine Division after I leave here. I can ask them if they would like to use one of their quad 50s mounted on a One Ton Truck to come down Hwy 1 next to the railroad track just before they start firing and hit them from behind. The Koreans have three of these trucks and are only ten minutes from the planned Viet Cong firing point for their mortars. I also have an agreement with the Navy for swift boats to support my PRU Tiger Scouts. I'm sure they can provide at least three boats to put Navy Seals into the water, and they will eliminate the threat from the sea." The Lt/Col answered telling Gilbert that the Airbase Commander would never approve and then added, "The USAF has Air Police and nothing can stop them from doing their job of protecting the Airbase."

Gilbert had served over twenty years as a USAF Tech/Sergeant before he took this job with the CIA. He knew they were good but this was like the Army Tankers that Gentle Ben had told him about. The Army Tank crews had been chasing a North Vietnamese Tank Battalion for years and had finally gotten them into a position where they had to fight. It would become the first tank to tank battle of the war. Gentle Ben and Helen his wife were flying at 4000 ft in a UH-1 watching the tanks maneuver. Helen kept telling him that they had to help but all he did was point out that the artillery had stopped firing, and even the USAF fighters were circling at 30000 feet waiting. The time was about 1600 hours and by dark the battle was over. The Army tanks began forming their wagon wheel defense circle by the light of the burning North Vietnamese tanks. Gentle Ben and Helen were in a hotel in Saigon having dinner with a South Vietnamese Government Minster Mr. Cat Lee and his family. This just wasn't the way that Gilbert would have stopped the enemy. He wouldn't have given them a chance to fire one mortar round.

A long two minutes passed while the Lt/Col spoke on the phone to the Base Commander and then the S-2 Officer turning back to Gilbert said, "Thanks for the information. We can handle it. I still need information on a small problem that is bothering me." Then he told Gilbert that the airbase was a restricted area and no USAF personnel were allowed off base except for a very few with special passes approved by the Base Commander. Somehow the airbase was being swamped with hard drugs, and the Air Police couldn't stop the people selling the drugs. USAF personnel were assigned to the base for one year tours and had to leave the country for R&R leave ever six months. The USAF only paid the airmen around sixty dollars per month in script, and the rest of their pay was sent home. It was like a prison that was locked down but the hard drugs still got into the base. Then the S-2 Officer said, "It must be the Viet Cong. Maybe you can catch one and torture him until he tells you how they are getting the drugs onto the base?"

Gilbert was shocked, just as he had been thinking that the S-2 Officer was his friend he discovered that the man thought he got his information just like the North Vietnamese did. Gilbert actually thought that the man knew what the boat rides at the party were for, but now knew that the USAF Command had no idea and didn't really care about what he did. They wanted information but didn't understand or care how many Vietnamese were killed getting it.

He had let his past experiences as a T/Sgt in the USAF influence his thinking, and in his line of work that could be a fatal mistake. Gentle Ben once told him that when he was in a firefight the first enemy bullets always seemed to hit his radio. He laughed as he said, "the last thing I need is someone sitting at a desk back in Saigon telling me what to do."

Gilbert left after telling him that he would check with his White Mice (Vietnamese Police) to see if they knew where the drugs were coming from. As he drove through the airbase he thought the Lt/Col is right, it does look like a prison, and then told Monk and his Interpreter, "I'll bet he doesn't know that "White Mice" is a nickname for the Vietnamese National Police." Then he asked, "Do you like Korean food?" They just looked at him nodding. No one can hear what's said in an open jeep approaching 70 MPH on a dirt road. They had no idea what he had asked them.

When they came to Hwy 1 Gilbert turned left and went flat out heading for the entrance to the Korean Compound. The guards heard the jeep coming and had opened the gate while waving them through. Gilbert hit the brakes and the jeep skidded a hundred feet before it could be turned towards the S-2 bunker. As he looked back the guards had closed the gate and were looking down the road with drawn weapons searching to see if the devil was really chasing him. It felt good to be in a safe place even if only for few hours. No need to look over your shoulder or worry about what was around the corner. Once again he left Monk and his Interpreter with the jeep as he walked into the S-2 shop. The Korean S-2 Captain greeted him and asked? "How about lunch?"

When he agreed the Captain phoned the Division Commander Colonel Lee and told him that Gilbert had agreed to have lunch with him. As they walked toward the headquarters' building the S-2 Officer said, "I've sent my sergeant to escort your men to the NCO Officers mess. The Colonel has a private dinning room, and we will meet him there." Ever since Gilbert entered his office they had been speaking and laughing in German. Both men were fast becoming friends but best of all they looked at things in much the same way. Follow the rules and beware of the dog that bites. In this case it meant always speak in English when in the presence of the Korean Marine Division Commander as he didn't understand German.

Lunch with the Division Commander turned into an hour and a half meal. The only problem was that there wasn't one dish that Gilbert could recognize. The dishes he ate were so spicy hot, that he still couldn't taste a thing the next day. Attending the meal were the Colonel's complete staff, and while eating Gilbert told them about the planned Viet Cong attack upon Tuy Hoa Airbase. The Division Commander asked if the USAF would like them to take care of the Viet Cong on the railroad track, and when Gilbert told them that the Air Force didn't want any help he laughed and said, "I understand he wants his soldiers to make the kill. Will they cut off their heads and mount them on poles along the road leading into the base?" "I don't think he has really thought about what is going to happen." Gilbert answered.

The Colonel then asked, "I understand that you have a speedboat. Do you have any water skis?" "I sure do." Gilbert answered. Then they got down to some serious planning. This coming Saturday Gilbert would bring his boat up the canal that ran from the South China Sea along the eastern edge of the Korean Firebase so the Division Commander could water ski on the canal. That Saturday when Gilbert arrived the concertina wire had been opened up and rolled back so the Colonel could walk barefooted to the canal. There were marines lining the banks spaced twenty feet apart acting as guards. Those that had their backs to the Compound faced the canal and on the other side they faced away guarding the canal. It appeared to Gilbert that the entire Marine Division was guarding the canal so the Colonel could ski up and down the waterway without worrying about snipers. When he fell off of the skis no one laughed, and when he offered Gilbert a turn he refused saying that he wasn't feeling well. Gilbert was actually a good accomplished water skier and smart enough not to let the Colonel find out. He needed the Marine Whitehorse Division and wasn't going to become a competitor in anything.

Gilbert had ridden over to the Korean Compound in the speedboat and while returning back to the PRU Compound had sat thinking about the USAF airbase's drug problem. He had his paid informers asking questions in the city, and had his PRU Tiger Scouts checking on all the Vietnamese who worked on the base. He even had the police stop and search everyone as they turned off of the railroad bridge toward the road leading to the airbase.

He knew that the drugs were being brought to Tuy Hoa City but no clue yet on how the drugs got into the airbase. He decided to drive out to the airbase and check to see on how they had made out with the sapper (terrorist) attack. Traffic over the railroad bridge was slow that morning and it was close to midmorning when he turned on to the side road toward the airbase. He had stopped to wait for someone to push a hand cart out of the way when he noticed a ten year old boy with a slingshot aiming at one of the guard towers. There was no one else on the road. The boy shot a rock at the tower and a guard used a slingshot to fire a rock wrapped in paper back at the boy.

He started letting the jeep move at a walk, slowly edging closer to the boy with the slingshot, hardly moving as they watch the drama unfolding before them. The boy unwrapped the rock and put the paper that was wrapped around it into his pocket. Then he took a package from his pocket and shot it over the fence of concertina wire to the base of the guard tower. An airman climbed down from the tower and retrieved the packet, placing it into his pocket. They had just witnessed how drugs were being smuggled into Tuy Hoa airbase. His Interpreter asked, "What are you going to do?" Gilbert said, "We are returning to the Province Interrogation Center (PIC)."

He needed time to think about how to help the USAF and not destroy his contacts with the S-2 Officer and the hospital. He had found out their reaction when he had told them about a forthcoming attack by the Viet Cong, and how they would not let the US Navy or Korean Marines help them. This could be worse if he just told them that it was their stupid plan that made the airbase almost like a prison, and that the Air Police were the smugglers trafficking in drugs.

This was one time he needed advice from Bill Colby. That afternoon he phoned Bill, and briefed him about the USAF Base Commander and the S-2 Officer's reaction to Gilbert notifying them of the Viet Cong attack. Before Bill could answer he also told about how the Lt/Col S-2 Officer had asked for his help and what he had discovered. There was dead silence on the phone for a minute and then Bill answered saying, "Don't do anything. I'll phone you tomorrow."

Terrorist attacks by the Viet Cong had slowed down, and the Province Interrogation Center (PIC) was working well without Gilbert being present. Most of the Viet Cong Communists (VC) was recruited to help the South Vietnamese government or being killed in firefights until the daytime curfew was no longer needed and even the night curfew had been shortened. The Dark Force still made occasional terrorist attacks but recruitment by the VC had all but stopped, and the Communist tax collectors had all been killed. Now it was mostly drugs, misinformation, and lies spread by the Communists that kept the police busy but that was only part of the political process. Gilbert had been sent to Tuy Hoa City by Colby to eliminate the threat of terrorist attacks, and now the government was once again in charge. Everyone liked sleeping without waiting for the next rocket or mortar round to hit.

The next morning Bill Colby phoned from Saigon and told Gilbert that the Province Officer In Charge (POIC) was going home to the states and another man would be taking his place. He was to take over the man's assets (informers) and continue working them. Then he said, "Joe," using his given name. "There will be a courier on the Air American plane tomorrow that is bringing instructions and orders for the POIC to return home. He will also be carrying an intelligence packet for the USAF S-2 Officer. In that packet will be instructions to the S-2 Officer to brief the new Province Officer in Charge (POIC) about how the USAF is solving their drug problem, and informing him that you have a copy of the USAF drug warning issued by MACV."

"The Commanding General of the Air Force in Saigon has sent him a letter stating that their Criminal Investigating Division (CID) has obtained information about possible drug involvement of Air Police on airbases in Vietnam." He then added, "The Paris Peace Talks have taken a turn against the South Vietnamese. We may win the war here and lose it with the Paris Peace Accord."

Then Colby said, "The courier is also bringing a packet for your eyes only about a special assignment starting with this phone call. His instructions are to identify you as the man accompanied by a Buddhist monk in orange robes. If there in any indication that anyone is expecting the packet he will not deliver it. He will however give you an envelope that you must sign for first. The envelope has instructions for you to turn over a complete list of members of the PRU Tiger Scouts and even a list of those killed in action (KIA) to MACV Headquarters."

The list will be added to a master list that would be exchanged for a list of American prisoners being held in the Hon Lo Prison (Hanoi Hilton) in Hanoi. Then Bill suggested that Gilbert might want to send Monk back to Cao Lanh before he made out the list." He didn't have to tell Gilbert what was going to happen to everyone on the list if the Americans abandoned Vietnam. Someone in the State Department must know that the Communists were going to treat them as traitors and kill them. Bill Colby couldn't be the only person who knew that the Communists were beaten, and it was only the Paris Peace Talks that kept the war alive.

Then Bill started to jump from one subject to another, and Gilbert thought that it must be really be tough for a man like Bill Colby knowing that the Americans could win every battle or firefight and stop the Communist Terrorists when ever they wanted yet still lose the war because some people wanted money more than honor. He told Gilbert that he would not need his interpreter in his new assignment. Then he added that there would be a DC-3 with no markings landing Black at Tuy Hoa Airbase at 1600 hours on Friday. The passengers would be traveling Black with a Lieutenant and twelve men. They would arrive after the Province Officer In Charge (POIC) had left Tuy Hoa, and that the PRU Special Forces Advisor or the ARVIN PRU Chief were not to know about this assignment.

He was to meet the plane with the PRU Pickup and house the men on the airbase. The USAF was not to know they were there. No one except Gilbert and the men themselves were to know what they were doing. A Cambodian Lieutenant Chga Bhun Youeca who spoke French, English, and Vietnamese would act as his interpreter. All other instructions would be in the packet that the courier was delivering to Gilbert. This was more like what he thought it would be like when he contracted to work for the CIA.

Then Bill said, "The airbase barracks area took five mortar rounds during the Viet Cong attack. One airman was killed and three wounded. The Air Force captured three mortars but the Viet Cong firing them escaped with no casualties. The Air Force killed two of the swimmers with only one aircraft slightly damaged. The Airbase Commander is being recommended to receive the Silver Star for Gallantry in action against an armed enemy of the United States. There was no mention of anyone telling them about the forthcoming attack."

Then Bill, using Gilbert's given name said, "Joe, I'm going to leave a sign that all Communists will understand, and historians will someday be able to interpret. Remember that this is the last war where the enemy will stand up and say, I'm the one. Your next station will be in Nha Trang. You've got about seven weeks to finish up here. Be careful."

Gilbert wasn't sure just what Bill Colby was talking about but he didn't need anyone to tell him to be careful. He was in his office at the Province Interrogation Center (PIC) when he took the phone call from Bill. After hanging up he sat for a few minutes thinking about what had and hadn't been said. Then he started opening his desk drawers and emptied them in his wastepaper basket for burning. He even destroyed the interview tapes of interrogations he had been studying. He only had four days before his guests arrived, and somehow he must protect as many PRU Tiger Scouts and their families from the Communists as he could. He couldn't save them all as that would surely be like signing a death warrant for everyone. This was not at all like he thought it would be working for the Company.

There was no one he could turn too. Gilbert learned to pray that day, and between praying and cursing he accomplished miracles in the next few days. He prayed for Bill Colby to have divine guidance for he sure didn't feel that he had. He first thought about asking his PRU Special Forces Advisor to help but remembered that no one could know or everyone would die. Monk was sitting at his usual spot watching and waiting to see if he was needed. Gilbert arose from his chair and walked out to the front part of his office where the two charts and the province map were hung. His secretary was at her desk and the interpreter was on the bench with his M-16 at his side. Monk had followed him out as he seemed preoccupied. Without speaking Gilbert picked up a grease pencil and started writing on the charts.

First he went to the map and drew a large circle around an area of Hwy 1 that included part of the western section of the USAF Airbase and the Korean Marine Compound. There was nothing but the circle in red crayon on the map. Then he went over to the chart titled Awards/KIA and wrote ten RVN Gallantry Cross w/Palm medals and ten Caskets. He didn't list any names or dates. Then he turned and spoke to his secretary, "Prepare a voucher for funds to pay the families for ten PRU Tiger Scouts killed in action (KIA). Don't forget to order the caskets and medals." She answered saying, "We have the medals on hand. How soon do we need the caskets delivered?" Without answering her question he said, "Have the vouchers for the funds prepared for my signature and ready when I return from the airbase."

Gilbert had already determined that he could remove ten men from the list being turned over to the Communists but who the ten would be he just wasn't sure. He knew that he couldn't just leave them off the list but to give them and their families a chance he had to list them as KIA. This wouldn't get through MACV unless money was drawn to pay the families. He was already on the Communists hit list so that wasn't anything to worry about. All they might do was increase the bounty on his head. If the Company put him in jail for buying caskets that weren't needed then at least he would be in good company.

All the Vietnamese PRU Tiger Scouts who saw the chart and heard what he had done began to tip their hat and bow towards him when they passed. Gilbert thought that this was a strange reaction. Monk got more protective if that was possible, and when he asked his interpreter why the soldiers were greeting him in this manor the man said, "Vietnamese do not fear death. He has given them a great honor by preparing a box for them if they are killed. By providing for their family he must truly be a great leader." Gilbert found it hard knowing that he was only going to save ten soldiers and their families but hundreds would die.

The funds and the caskets were available the following day. Gilbert called his Interpreter into his office and said, "I have been given a new assignment, and no one will take my place. Your job no longer exists. You have done me and your country a great service. Please accept this Gallantry Cross w/Palm as a token of appreciation. Have your wife come and get your death benefit pay. Remember don't come with her. Tonight move out of your hutch, and use a different name as you move about." Then he walked out to the front office where Gilbert entered the man's name and date on the KIA Chart.

As he filled out the chart Gilbert told his secretary that she was to prepare a list of those men KIA for his signature. After the list was completed she was to send the list of all remaining PRU Tiger Scouts to MACV. Then he said, "If you misspell some of the names it can't be helped. Just be sure that the names on the pay vouchers are spelled with the same mistakes." His secretary was a well educated woman and never made spelling mistakes especially when it came to Vietnamese names. She just smiled and answered, "What if I misspelled all the names?" "I wouldn't know the difference as I can't read Vietnamese anyway," Gilbert answered.

When Friday came all the names had been filed in but one. Monk had taken to sleeping in the front entrance of the French Mansion at night. It was hard for Gilbert to even go to the bathroom alone. Everyone seemed occupied in protecting him from the Viet Cong even though they were no longer a threat in the city. He had the packet Bill had sent him and had read the instructions three times and would read them again before burning the papers. Gilbert had taken to driving the pickup truck so no one was surprised as they left the house in the truck for the Tuy Hoa MACV Headquarters.

As they left the Compound Gilbert asked, "Monk, have you brought all your things? You will not be coming back." "When we leave a place I never plan upon returning. This is no difference," Monk answered. Gilbert drove into the National Police Compound and stopped at his office. "Wait in the truck," he said. Going into the office he went directly to the KIA Chart and added Monk's name with today's date. "Give me his family's death benefits." Gilbert said. Then he told his secretary that Monk had no family in Tuy Hoa, and he would see it was delivered to his people in Cao Lanh.

After he received the money he stepped out and called for Monk to come into the office. Then Gilbert said, "Change into your Orange Robe. Leave your M-16 and uniforms on the chair. You will be leaving for Cao Lanh on Air America today. I'll miss you but you can't go with me on my next assignment." As they left his office Monk just looked at the KIA chart and grinned. Their next stop was at the Tuy Hoa MACV Compound where they had a phone link to Saigon. Monk stayed in the truck while Gilbert called Saigon Air America Operations for a ride to Cao Lanh for his bodyguard. All he had to tell them was that "Hotel Man" needed a plane at location #103 and its destination. He was informed that it would arrive in two hours.

Then he went into the Province Officer In Charge (POIC) office. "I understand that you have the names of your Viet Cong informers that I'm to run while you are gone," Gilbert said. "I've got one Viet Cong that is a key man working in a North Vietnamese Army Battalion Headquarters and supply point located just west of Hwy 1 and south of Hwy 7 near the Song Ba River. They even have a medical surgical hospital assigned to them. We've kept the location secret from the USAF and the Korean Marine Whitehorse Division as he is a key man there. He meets with me every two weeks and after he furnishes me with information on what they are doing or planning I pay him," the POIC said. Then he went on to say, "He's a very intelligent man and educated. Don't let the Koreans find out as he gives us a lot of information about the Paris Peace Talks. I've just met with him, and it should be about two weeks before he will want more money."

No wonder Colby was sending him home Gilbert thought to himself. This POIC was helping to hide a complete North Vietnamese Army Battalion and supply dump next door to the USAF and Korean Compound just to protect a spy. There was no way Gilbert could follow the POIC instructions with this time bomb waiting to explode and destroy Colby's plan. "Have you had the Viet Cong authenticated?" Gilbert asked. "Yes, I did. I had him Black Boxed (Polygraph Tested) and his information has checked out," the POIC answered. Then he said that the Viet Cong would leave a signal by making a mark on a light pole just before the turn from Beach Road onto Hwy 1 beside the Catholic Church. They would meet the following day at noon in a small room next to the church's confessional booths.

"If you are ready I'll give you a ride with my Buddhist Priest to the airstrip," Gilbert offered. Monk got in the back with the baggage while the POIC rode in front with Gilbert. The Air America plane had just arrived and Monk unloaded the baggage from the truck. Then he turned and bowed low with the hands pressed together in the Buddhist manor. Gilbert returned the greeting in the same manor and said, "Show me your begging bowl." When it was presented Gilbert placed an envelope with a years pay, the RVN Gallantry Cross w/Palm medal, and a Joss stick in it. Then he said, "Please burn this for me in your Temple." The priest held out his hand to Gilbert, and as he shook it, said in perfect English, "I shall always remember you."

This from a man who had never spoken a word of English since Gilbert knew him. It all started to make sense. Monk was more than a bodyguard. He was Gilbert's protector.

When he sat outside of restaurants with his bowl it was so the Viet Cong would not attack. He was part of the Buddhist Group of Priests who were against the Communist. Every time Gilbert had been ambushed Monk wasn't with him, and the VC really wanted him out of the way. Gilbert then remembered that at night if Monk stayed in his house there were no mortar attacks. It all started to make sense, and now he understood why Monk treated everyone the same. This was also why everyone in the Province Interrogation Center (PIC) were really afraid when Gilbert was giving them hell for torturing a prisoner as Monk would kill them in a moment if Gilbert gave the sign. Even Colby must have guessed that Gilbert had the approval of the Buddhist.

As soon as the currier delivered his packet Gilbert left for the Tuy Hoa Airbase as he was now on a tight schedule as there was a DC-3 plane due in shortly with the group of twelve Cambodians who he would be training for the next two weeks. His instructions were simple. Meet the plane landing at Tuy Hoa Airbase, house 12 Cambodians and one officer, feed them from an Air Force mess hall, train and equip them for snatch and grab raids in Cambodia with out letting the USAF find out. Then do it again with another twelve men. The packet for his eyes only had left out two things. One was how to do this little deed, and the other was how long this was to go on.

Now Gilbert had been in the USAF for twenty years as a T/Sgt and knew how an airbase worked. To house the Cambodians he went to the Base Engineer and checked on all vacant buildings. The Engineers keep a record of all buildings by a building number, and each building that is occupied must be signed for by the using squadron. Gilbert was only interested in those that weren't in use. He found three and picked the one near the beach close to the officers club. It was one of the barracks that had been emptied when they were preparing for the attack from the sappers (terrorist) the week before.

Gilbert used his walk on water CID Card and signed for the building. It was a one story wooden barracks located with the sea on the east and across the street from the officer's club. The Lieutenant could live in a single room in the barracks and use its latrine. The soldiers could use the floor for sleeping, eating, and classrooms. The barrack's had one latrine (bathroom) with electric and running water. Gilbert planned on using the twenty-four hour mess hall of the Air Police to provide carryout meals to the men in the barracks. There would be no trouble with washing dishes as all military organizations including the USAF used paper plates, cups, and plastic eating utensils. LT Chga Bhun Youeca would eat lunch and dinner with Gilbert at the officer's club. Breakfast would have to be American "C" rations for everyone.

It was a good plan except for one small but important item. The USAF ordered its food supplies based on the number of meals served to save money on food and accounting costs. They did this by counting the paper plates, and the supply depot recorded the number of plates as they were sent to the different mess halls. This was checked against the tallies of the individuals on separate rations besides being used to determine the number of meals served. It was quick method of determining if someone was stealing food. It was also a second check against the personnel records to determine the number of airmen on leave or on R&R at anytime.

This number was sent to the accounting department of the finance center to be matched with the number of airmen on the base. Due to this oversight a few days later before the graduation of the first batch of Cambodian trainees the Base Commander accompanied by the Air Police Commander and a truck load of Air Police were able to find the barracks where Gilbert was training the Cambodian's Snatch and Grab team.

When the DC-3 arrived bringing in Black the Cambodian men to be trained as snatch and grab soldiers Gilbert met the plane. His pickup truck was in the parking area near the Operation Building. After the plane was parked the crew chief opened the door. It was actually one half of a cargo door large enough for a jeep, and then he installed the boarding steps in the floor.

While working as a parachute rigger in the Air Force for twenty years Gilbert was familiar with the C-47's of WW-II. As he approached the plane he noticed that this plane had the plastic navigation dome on top just behind the cockpit and was still wearing the old Army drab paint. The windows also had the small round gun ports in the center of each one but it was the uniform of the crew chief that caught his attention. It wasn't Air America's white shirt and dark paints. It was a one piece flight suit and over the left shirt pocket was the insignia of Continental Air Lines.

He climbed the steps and walked into the cabin. The floor was of metal with tie downs, and the seats were canvas benches running along each side. There was even a static line running overhead down the center for parachutists to connect their ripcords. Everything looked bright and new ready to be used. The soldiers and their Lieutenant were still sitting waiting for someone to tell them that this was the place. Gilbert just waved them to stay seated as he proceeded up to the cockpit area. There was a navigation table on the left and across from it was a radio station. The navigation station had an ARN 6 radio compass, a magnetic compass, an altimeter, and airspeed indicator, a drift gage (sight), a sextant in its case, and a loran "C" radio. The only thing out of place was a low pressure oxygen mask lying on the table. As he moved closer to the pilots he noticed they also wore the same jump suite as the crew chief with the same insignia. The fold down seat for the flight engineer had an oxygen mask laying on it. As he spoke to the pilots he saw that both had masks lying on the deck beside their seats. He also realized that these were not USAF oxygen masks but full face masks like firemen wore when fighting building fires. When Gilbert first boarded it was like walking back into the 1940 and WW-II. Now he was looking at an instrument panel with ILS, UHF Radios, HF radios, and even intelligence coded receptacles that were used to the prevent anyone from tuning into their radio frequencies.

There were two switches with red covers just below the throttles. On one cover it read "JETO" (jet assist) and on the other the cover read "TOW" (a release for gliders). When he first entered the C-47 he had seen that the tail part of the fuselage was blunt with a hitch for towing gliders sticking out. There was a sweet smell of something in the cockpit, and he suddenly recognized what it was. He had heard rumors that the CIA was moving cannabis on their planes. He was standing in one of the planes that moved in the Black and also carried drugs. He knew that Vietnamese Airlines flew in and out of Vietnam to Thailand, Cambodia, and Laos. Everyone thought that Air America was the Company Airline and it was, but Gilbert hadn't realized until now that the Company had contracts with other airlines.

When carrying bales of cannabis the dust from the drug would make a pilot drunk or even fall asleep. It would become so strong in a closed plane cabin that the crew's hands would shake but they felt good. You didn't have to smoke the stuff to get where you don't care if you missed the runway or landed fifty feet in the air. Wearing the masks breathing clean air was the only way a crew could operate this plane when it was carrying cannabis. This was one of the cleanest airplanes Gilbert had every seen. He was fascinated with the cockpit and would have like to ask questions but the Captain pointed back at the passengers with his thumb and asked, "These yours?" Gilbert answered, "Yep, Inform the tower that you have to check an engine before taking off. Ask permission to taxi to the maintenance area. Tell them you are having your crew check it out. Park so they can see your plane but not the door. I'll meet you there and take your passengers." Gilbert hurried off the plane, jumped into his pickup, and drove to the end of the ramp to wait for the DC-3. As he watched he couldn't help thinking 'what an antique". It looked as if it was on its last legs outside but inside it was a dream.

After picking up the Lieutenant and his men he took them to the barracks where he instructed them that no one was to go outside unless he was present. Then while the Lieutenant got them settled in he took three men and drove across the base to the Company's warehouse. He and the Province Officer In Charge (POIC) had the only keys. In it there were uniforms, camping equipment, squirt radios, weapons of all types, claymore mines, and everything needed to outfit Province Recon Units. His special instructions were to train these soldiers in the use of the field squirt radios and equip them for a snatch and grab raids into Cambodia. His instructions were that he had two weeks to get them ready. They had been moved twice before to different bases for special training, and now Gilbert was to put the finish touches on before sending them out. The plane picking them up would also deliver another group; The Lieutenant would stay and help him get each team ready.

On the way back from the warehouse Gilbert stopped at the Air Police mess hall and took thirteen dinners to go. They were put in paper plates and containers and was better food than most of the Cambodians had eaten since the war started. As they had been traveling all day this was going to be their lunch and dinner. It was 1400 hours and Gilbert was quite pleased with how smooth things were going.

A little over a hundred miles away Gentle Ben had just delivered a water trailer to an ARVIN Company close to Loc Ninh that had been with out water for two days. As the Chinook lifted off from Loc Ninh he received a radio call from Saigon Special Ops (TOC) to proceed to Quan Loi and pick up an internal load for immediate delivery to Gilbert at Tuy Hoa Airbase. He was told that the load was classified and must be delivered prior to 1600 hours. It was on the artillery log pad located at the supply depot on Quan Loi. This was where he picked up the gun crews, stingers (105 mm tube) and ammunition for artillery raids into Cambodia so he wasn't surprised at being instructed to pick up a classified internal load from this log pad. It was only ten minutes to the log pad and allowing fifteen minutes for loading with an hour flight to Tuy Hoa he would even have time for a hot refueling stop.

With hot mikes the crew could talk to each other and yet hear what the instructions were. This saved Gentle Ben from having to repeat what they were going to do. While he was talking to special operations or to other aircraft he turned off his intercom so he couldn't hear his crew but they could still hear him and the party he was talking to. When he came back on the intercom he said, "It looks like we are in luck this time. We'll get back to Bear Cat in time for supper. Helen's sure going to be surprised to see me home before midnight."

The Artillery Pad was a large pin-a-prime (oil to control dust) covered field. When they arrived there were three other CH-47 Chinooks helicopters picking up or delivering sling loads. When they landed the small tower acknowledged them, and then announced that their load was on its way. They had parked to one side of the log pad to wait. The load was about sixteen cardboard boxes stacked on an Army Mule vehicle. They were stacked higher than the drivers head, and he was in a hurry. This vehicle was developed as an ammunition carrier for the infantry. It was about 8 feet long by four feet wide and can be driven by a soldier walking behind it or sitting on a small seat in front. The maximum speed was like a golf cart just faster than a person running.

As Gentle Ben waited they could see the Army Mule coming full speed from near the base of the tower directly across the log pad. The helicopter crew had no idea what it was carrying but Chinooks develop a wind from their blades of almost two hundred miles per hour even while on the ground in flight idle. The first thing they noticed was that the boxes were not tied down, and if they were full of ammunition this was no problem. Gentle Ben said, "Flight Engineer, Load the mule and tie it down. We will use it to unload and bring it back tomorrow." Just then it came within the down draft of the forward blades. The first case started to come apart and paper plates started flying. The driver started grabbing at the plates. The vehicle turned sideways before stopping. All the boxes came apart and ten thousand plates went flying.

The Chinook has two engines mounted on its top near the aft rotor blades. If the plates blocked the engine intakes they could be damaged. All the other helicopters in the area pulled full power and left trying to get away from the flying paper plates. Gentle Ben couldn't leave as they could no longer see the Army Mule due to the flying plates. He pulled the helicopter's nose up so the wind from the forward blades would keep the plates away from the engines. With all the helicopters pulling full power it was like four small tornados pulling at the paper plates. After about five minutes the crew could see the Army Mule and its driver. He was standing looking at an empty Mule holding a paper plate in each hand. The Flight Engineer who was standing on the rear ramp connected to the helicopter by an intercom radio cord called out, "There are paper plates flying over the top of us. I can't see the sky. Everything is white." From the cockpit the pilots were now able to see the tower, and the rest of the log pad. There were no paper plates on the Army Mule or the log pad, they were all still in the air flying. It took another ten minutes before all the plates had flown away, and they were cleared for take off.

They climbed to four thousand feet, and Gentle Ben then radioed Saigon Special Ops and said, "Unable to load special cargo. Please advise interested parties that they flew away, over." He had to repeat the message three times before they finely came back and said, "Cleared from mission. Return to 1st Cavalry Control, Out."

CHAPTER FIVE

It was another bright beautiful morning in the Garden of Eden, the kind of day where it seemed that nothing could go wrong. Gilbert left the French Mansion for a stop at his office in the National Police Compound where he would sign papers and show his face before heading to the USAF Airbase to start the first day of instruction for the Cambodian soldiers in the use of the CIA field equipment. They would need all the help he could give them for their mission of watching and reporting on the type of traffic using the Ho Chi Minh Trail. Yesterday he had issued M-16s and one M-79 plus machetes and other field equipment. The company only issued defensive weapons for the Province Recon Units (PRU Tiger Scouts) or the snatch and grab teams. The US Army was the ones who were equipped and trained to defeat the Communists in any field battle. Today he would start training the Cambodians on how to use the squirt radios, weapons, and other equipment. He planned to have them work on their radio procedures every day for the entire two week training period.

Gilbert's schedule was to arrive at the Cambodians barrack at 1000 hours each day after Lieutenant Youeca had them ready for class work. At noon he and the Lieutenant would drive to the Air Police mess hall and pick up their lunches. While the men ate they would have their lunch at the officers' mess, and then more classroom work in the afternoon. At 1600 hours he would allow them to go outside on the beach side of the barrack away from the view of the base. They would practice dry firing their weapons and exercising with the M-16. He had borrowed a volleyball net from the USAF base athletic department, and was beginning to teach them the game. The soldiers would be allowed to play for an hour every evening before dark. After the men had finished playing volleyball they went to the showers while the Lieutenant and Gilbert went to pickup the trainees evening meal from the mess hall. While the men ate they would go to the Officers' club for their evening meal. The plan worked very well for the first two days, and then the results of Gentle Ben's disastrous loss of the flying paper plates caused the Airbase Commander to start looking for who was getting free meals.

It was just before noon, and Gilbert had everyone outside taking a break from the hot classroom. The Cambodian soldiers were all outside with their weapons stacked choosing sides for a volleyball game when the Base Commander showed up with his Air Police. With hands on his hips he asked, "Who the hell are you?" All the time he was looking directly at Gilbert, and his next question was, "what are you doing on my base?" Gilbert answered him in the same tone, "Colonel, you don't have the need to know. If you wait a couple of hours there will be someone from MACV in Saigon to brief you." Gilbert then sent the soldiers into the barracks with their weapons while he went to the Officers' Club to use their phone to call Bill Colby. He returned to the barracks and informed the Cambodian soldiers that they would practice eating American Army "C" rations for lunch.

The Base Commander had his Air Police surround the barracks to keep the Cambodian soldiers from escaping. This upset Gilbert a little so he showed him his CIA walk on water ID Card, and suggested that the Airbase S-2 Officer could verify who he was, and how he had assisted the Air Force in the sapper (terrorist) attack by the Viet Cong. This must have got the Base Commander's attention for he calmed down a lot. Two hours later Gilbert was notified by an air policeman that a USAF Lear Jet was inbound with a General Officer aboard who wanted him to meet the plane when it landed.

The same message was also sent to the Base Commander, and when Gilbert showed up at the Operations building the Base Commander was already there waiting. They didn't chat or trade stories but Gilbert waited out side of the building while the Colonel waited with his Air Police Commander and the Operations Officer inside. After the Lear Jet landed and was taxing into Base Operations the Commander stepped outside to join Gilbert.

After the plane's engines were shut down a USAF General stepped out and upon seeing Gilbert walked over greeting him while saying, "How are you Colonel?" Gilbert answered, "I'm fine, and this is Colonel Jones the Base Commander. He has asked to be briefed on what I'm doing here."

The General turned to the Base Commander and said, "Colonel, let's go to the Officers' Club. I understand it's near to Gilbert's class room. We can relax while we talk." When they arrived the General noticed that Air Police were surrounding the barrack where the Cambodians soldiers were staying. The General suggested that they would be better used to move everyone out of the Officers' Club and see that no one entered while they were having a little talk. After the building was secure, the General briefed the

Colonel on what Gilbert was doing and the support that was needed. It was a good briefing, and it was also the first time that Gilbert knew what the Cambodian soldiers were going to do. Colby had only told Gilbert what he wanted done, and then let Gilbert decide on how to do it.

At the end of the very secret briefing the General asked the Base Commander, "Can you live with this? If you can't just say so, and you can return with me to Saigon." Then he said, "I want you to help Gilbert's mission in any way you can." Then turning to Gilbert said, "You stay and get on with your classes. The Colonel and I can visit on the way to my plane. If you need anything just let me know." Gilbert had sat through the briefing saying nothing and pretending that he had heard all this before. He did a lot of nodding and agreeing to some of the finer points. He wondered to himself if Colby had told the General that he was a Colonel. If the General had addressed him as Sergeant he would have still just nodded. In the days to come he would continually be amazed at what Colonels would do to help a fellow officer.

During the meeting the General laughed and speaking to Gilbert said, "If that Warrant Office Pilot of yours hadn't blown ten thousand paper plates all over Quan Loi I wouldn't have got to meet you." Then he told them what had happened, and then added that Special Operations was getting about a dozen letters a day from commanders complaining about what Gentle Ben had done. Then he said, "I've got a bet on with General Abrams that this will bring more letters from commanders than the time he dropped all of the 1st Cavalry Division Artillery Battalion' s laundry on the top of the empty artillery casings thirty foot high stack at Quan Loi."

Then he went on saying, "That time one of the company commanders on firebase Burt declared an emergency, and he went in under fire to help. Then the Company Commander personally hooked up the sling with dirty laundry without his permission. It seems as if the Cavalry Troopers wouldn't do it, and kept telling their Company Commander that it was Gentle Ben's helicopter. The Company Commander was new in country and didn't know that Gentle Ben never carried beer, hard water (ice), slings, or laundry until all the firebases had ammunition for the night. The only exception was when a soldier was going home or on R&R he would make a pickup day or night, enemy fire or not. You know he got his experience flying with the USAF for twenty years." He broke up the meeting still laughing, and Gilbert still didn't know how ten thousand flying paper plates almost screwed up his training of the Cambodian soldiers.

The next day when Gilbert drove by the Catholic Church in Tuy Hoa he noticed that there was the prearranged signal, a mark on a elect power post, from the Viet Cong informer (spy) indicating that he wanted to meet with him in the Catholic Church. The agreement was for Gilbert to meet him the following day after the signal was posted in a small room off of the chapel in the church. Only a week had passed since their last meeting, and this worried Gilbert. Maybe the man hadn't removed his mark from the last meeting or it could be a setup for an ambush. What ever the reason it wasn't normal as he usually waited at least two weeks before wanting more money. Gilbert decided to go to the meeting site, as per their agreement, but this was one time he wouldn't have Monk with him, and he really felt alone.

On the following day at 0900 hours Gilbert parked the PRU pickup he was driving, and walked into the Church like he was there for spiritual guidance. He walked up the center isle of the chapel and kneeled in front of the statue of Christ. A Catholic priest saw him, nodded, and continued lighting candles in front of the alter. Gilbert crossed himself, and then arose proceeding towards the confessional booths. In the darkened chapel the priest couldn't see the small 25 caliber Browning automatic pistol he was carrying in his right hand. While his head didn't move his eyes were darting every which way looking for any sign of an ambush.

As Gilbert came up to the door of the room where he was to meet the Viet Cong Informer everything seemed normal, the door was half open, so he slipped inside with out touching it. The Informer or Spy was across the room where sunlight from a small high window lighted the room. The Spy was alone and was obviously waiting for him. Gilbert moved the pistol to his left hand and with his right hand shook the man's hand while they greeted each other as friends do. It was the same man he had met before but somehow he looked different. The man stood straighter, shoulders back, and spoke with more authority. The Spy said, "I'm not going to work for you anymore." There was a table in the room with two small chairs so Gilbert pushed the door closed with his foot and sat down indicating the other chair for the Informer. Gilbert was now in his interrogation mode. He had sat down so he could hold his pistol in his lap, and no matter what happened Gilbert was close enough to the VC to be sure that if anything went wrong the Spy (Informer) would be killed. It's hard to miss a target seated across a table from you even with a small 25 cal automatic pistol, and Gilbert was an excellent shot.

The Spy sat down across from Gilbert and he said, "I'm sorry to hear that you and I won't be meeting again. I've brought you money but I need some information for my report." The man smiled and answered, "I have word from the Paris Peace Talks that the Americans have agreed to furnish us the names of the traitors who are helping the US Army and South Vietnamese government." "Have you any thing that I can show my Province Officer In Charge (POIC) that this is true?" Gilbert asked. The Informer had only met Gilbert once before and had no way of knowing how much authority he had. So when Gilbert acted as if he reported to the Province Officer In Charge (POIC) he became arrogant and started speaking down to Gilbert. Now Gilbert was a true interrogator, and nothing he liked better was when a subject believed that he was smarter than the interrogator.

Now Gilbert had been told by the Province Officer In Charge (POIC), who had located this Informer, about how the man had been authenticated and how the results of the information he furnished was checked. Gilbert had also seen the man's tape from the Black Box (Polygraph Test) he had taken. Gilbert himself didn't use the Black Box as he thought that it could be fooled, and just didn't trust its findings.

The Spy held out sheets of paper for Gilbert to look at. They were copies of the PRU Tiger Scout rosters he had signed the day before, and sent by courier to Saigon MACV. He looked the lists over, and saw that the names of the ten PRU Tiger Scouts that he had reported as killed in action (KIA) were on the list but his Secretary's name was missing. Gilbert was also sure that none of the other names were correct, and that their addresses were also misleading. Upon seeing the list Gilbert acted startled and said, "This is a list of all of our Province Recon Units (PRU) Tiger Scouts and their addresses. Where did you get it?" The man looked smug and asked, "You know these men?" "Well Maybe," Gilbert stammered. Later Gilbert would write in his Journal, "I'm a good actor. Maybe I should get a job in the movies."

Then the Spy said, "With this list the men in my unit will be able to find and kill them after you Americans leave." Gilbert asked, "Are you a Viet Cong Officer?" He answered, "I'm a North Vietnam Intelligent Officer. My Battalion is a regular North Vietnamese HQ & HQ for this sector. We no longer need your help. I must go now. My General has invited me to visit him for lunch. This will give me time to send in your confirmation of the list to our negotiators in Paris."

Gilbert wrote in his Journal that evening, "I couldn't let him just walk out of the church as the North Vietnamese had accepted the list as correct, and only this man or his associates would know it was false if they checked on each individual PRU Tiger Scout. If I had killed the spy in the Catholic Church this would have brought attention on how the North Vietnamese had used our secrecy to hide their

Camp from the USAF and the Korean Whitehorse Marine Division for the past two years. I can't tell anyone but I have to do something."

After giving the North Vietnamese Intelligence Officer the money that he had brought to pay him Gilbert thanked him saying, "I will inform my POIC about how you have helped me." The spy was all smiles as he counted the money, and he actually boasted that there were people in the United States who were helping them win the war. He was trying to impress Gilbert when he said, "You Americans can't win. Even now one of your Navy Officers is in Paris helping us prepare the Paris Peace Accord. Someday he is going to be a great leader in your country with the Communist Party's help." Gilbert just smiled again and shook his hand while the man actually strutted as he walked out of the church. Gilbert went out and drove off in no apparent hurry. He turned left onto Hwy 1 and headed to the USAF Airbase thinking about how he could solve the problem about having a North Vietnamese Army Camp, a double agent (Vietnamese spy), and a Viet Cong headquarters on their door step with out alerting the Communists of what was going to happen next.

He looked at his watch and thought to himself. The Informer had to go west on Hwy 7 for about five miles before crossing the Song Ba river to enter the Headquarters of the North Vietnamese Camp. The North Vietnamese Intelligence Officer would need at lest two hours to get to Camp, and prepare his report to Paris and Hanoi. If he had to get permission and have lunch with his Commanding General that should take another hour. Gilbert needed to give him at lest three hours before he informed Colonel Lee the Commanding Officer of the Korean Marine Whitehorse Division of the location of the enemy's camp. If he was going to save his PRU Tiger Scouts and their families' lives he had to give the North Vietnamese couriers time to get out of the hell that the Koreans were surly going to deliver to that enemy camp.

When he arrived at the classroom in the Cambodian Barrack he parked at the Officers' Club and walked across the street to the barrack. The men were already hard at work field stripping their M-16s while blindfolded. The field squirt radios were lying on a blanket on the floor and would be the next item that they would work with. All the time they would be blindfolded. Their Lieutenant had been watching the soldiers work, and when Gilbert walked in he approached him nodding toward the door.

As they walked out he turned to Gilbert and asked, "Do you remember the Vietnamese Nurse assistant at the hospital?" Upon Gilbert saying he did the Lieutenant continued, "She is going to eat lunch with me in my room. I promised her American "C" rations. I was hoping that the men could play volleyball and eat on the beach." Gilbert laughed and answered, "We will go and get the men's lunch now while they are working. You take over for the rest of the day as I will be at the Korean Compound and will be out of touch. I'll bet she is expecting more than "C" rations for lunch."

After they picked up the noon meals from the Air Police mess hall Gilbert went back to the Officers' Club and continued waiting for the Base Commander to arrive for lunch. Ever since the visit from the USAF General the

Base Commander just happened to met him every place he went on the airbase, be it the AFEX (air force store), or the Officers' Club. The Colonel had become a real help in getting things accomplished. Now Gilbert wondered if he would go for this next request. Just then the Base Commander pulled up in his jeep and walked into the Club. Without being asked he joined Gilbert at his table.

Gilbert said, "Colonel, I need a favor. These men I'm training have never seen a fighter bomber mission with a spotter plane working together. Is there any way you can have two flights heading on a live mission take off and circle about five miles west of the airfield at twenty thousand feet until I give them a target from a CC (Command & Control) helicopter?" The Colonel called over to one of the squadron commanders who had just came in and asked him to join them. Gilbert repeated his request. The Squadron Commander answered, "We can do better than that. If you can use our training flights we can provide you with two flights of four planes at around 1400 hours." Gilbert asked, "Will they have live bombs and plenty of ammunition?" and then, "How long can they stay on station?"

The Colonel answered, "They can remain on station for at least an hour or a little more. They will wait until you scramble them from the CC Helicopter. It will be good training for the new pilots. Will the target you pick be close enough for your trainees to see the bombs hit?" Gilbert answered, "Yes they will." Then he asked for the control sign of the FAX SHIP and frequencies that he would need. Then he said, "I have to go to the Whitehorse Marine Division and borrow their UH-1 CC Helicopter. I'll call you and test the radios in about an hour." As he left the table he overheard the Squadron Commander ask how long he should wait for Gilbert's call, and the Colonel answering. "All day if need be. Don't worry this is only a training mission, and the waiting will be good practice for your pilots."

An hour and a half had passed since Gilbert had left the Catholic Church and it would take him at lest fifteen minutes to reach the Marine Compound from the USAF Officers' Club. As he drove through the gate of the Korean Compound their military police directed him to the parking area at the S-2 bunker (office). The S-2 Officer immediately started speaking in German and Gilbert said, "I need to speak to Colonel Lee. I've got some good news but it won't last long." The Captain S-2 Officer could tell from Gilbert's continuous looking at his watch that something serious was bothering him. "Come with me. He is just finishing lunch with his Staff." Everyone with the Colonel was standing and preparing to leave the room as they entered. The Captain said something to the Colonel in Korean while nodding toward Gilbert. Everyone stopped and turned facing the Colonel as he greeted Gilbert. "Everyone speak English so I can understand you. German will not be spoken in my presence," He said as he glared at his S-2 Officer. "How may I help you?" He asked.

"Would you like the location of the North Vietnamese HQ & HQ Battalion's Camp at Tuy Hoa?" asked Gilbert as he handed him a rolled up air chart of the area. He had gotten it from the airbase operation's office and had entered the coordinates of the North Vietnamese Camp that he had gotten out of the old Province Officer In Charge (POIC) files. The Colonel forgot that he had ordered that everyone was to speak English and began shouting orders in Korean. At the same time he grabbed the table cloth pulling it and everything on it falling onto the floor. Laying the map on the table he and his officer started studying it with some making notes.

Without looking up the Colonel asked, "How old is this information?" Gilbert answered, "Two Hours." Now the Colonel looked up and asked, "How long do you think that this information will be stable?" Once again Gilbert answered, "One to four hours." The Colonel then asked, "Why didn't you bring me this when you first received it?" Gilbert answered truthfully, "We had a North Vietnamese informer in their S-2 but he just turned sour. I was going to kill him myself two hours ago, but I thought that I have a friend who could receive a star for destroying a complete North Vietnamese Camp. What was I to do? Help a friend or get revenge. Here I am." The Colonel went back to the map calling out orders.

Gilbert just stood and watched. No one gave him a second glance. Before his eyes somehow everyone was able to change into full combat gear. The Colonel had been given a bullet proof vest and some one had even handed Gilbert one with a helmet. He spoke to his friend the S-2 Captain saying, "I've made arrangement for the Navy to protect your Compound's southeast perimeter by patrol boats on the canal. Tell the Colonel that the US Army is stopping all traffic on Hwy 1 and Hwy 7 starting in one hour. My PRU Tiger Scouts will sweep the city and every thing east of the railroad track from Tuy An to Phu Khe. All prisoners will be turned over to your guards at the gate of your compound. The Navy will take care of the Song Ba River up to An Khe, and the USAF will have fighter bombers scrambled upon my order. They will be on a training exercise and will only hit targets that you provide. Everything from Nui Vang Phu Mountain north to Hwy 7 and east of Hwy 1 is now under the Colonel's control.

The Colonel was headed out of the door as the S-2 Officer and Gilbert hurried to catch up. As they left the building there was a UH-1H waiting. It was the Colonels Command and Control helicopter (CC). There were two pilots, two door gunners, and one radio operator already aboard. The Colonel sat in the center facing forward, the radio operator sat facing back toward the left door gunner. The radios were stacked to the operators left in front of the Colonel. Facing forward beside the Colonel was a Lt/Col Tactical Operations Officer who radioed the Colonels commands to the Marine Battalion Commanders. On the Colonel's right sat Gilbert. Normally this was the S-2 Officer's position but not today. The S-2 Officer sat facing Gilbert and the other door gunner.

The helicopter's blades were turning as they boarded, and the sky was full of helicopters circling or loading waiting for take-off. It looked as if a complete Army Aviation Battalion of slicks with a company of Cobra gun ships were there. Gilbert looked at his watch and a little over thirty minutes had passed. The complete Korean Whitehorse Marine Division would be soon be airborne, and attacking the North Vietnamese Army. Gilbert had planned on giving the North Vietnamese Intelligence Officer three hours to get his couriers away before the attack began. It was going to be close. He just hadn't realized how bad the Korean Marine Division Commander wanted to destroy that North Vietnamese Camp. The S-2 Officer leaned forward, and told Gilbert that the Colonel had been searching for this Camp for two years. He had been promised his star when he destroyed it.

The Colonel asked over the intercom, "Why are you always looking at your watch?" Gilbert answered, "I gave the informer some false information for the Paris Peace Talks, and I estimated he would need three hours to get his couriers out of the kill zone." "What time will that be?" the Colonel asked. Gilbert just held up two fingers. The Colonel never answered but told his Operation's Officer to have all the helicopters head south toward Nui Vong Phu Mountain. Its height is 6729 feet, and they should all be able to see it. Then he said, "I'll tell you when they should turn so that our attack will be out of the south with the sun to our right. They will never guess that they are the target." Then turning back to Gilbert he said, "Just make sure your Air Force guys don't spook them."

Gilbert gave the radio frequency of the Navy Operations to the radio operator and the frequency of the Air Force spotter plane. He then contacted the Navy and requested that they send Swift Boats up the Song Ba River as far as Cheo Reo. They were to stop and search all boats as they started back down toward Tuy Hoa. All troops on the south side of the river were considered to be friendly unless they were fired upon. He told them that they were supporting a Korean Marine Division who was maneuvering in the area.

While he was doing his thing with the Navy the Colonel had ordered the Division to turn south and began the attack. Gilbert radioed the USAF and scrambled the fighter bombers. He then called the Air Base Commander and told him that the US Navy was patrolling the river all the way to An Khe. Also that the Korean Marine Division was conducting a training exercise to the south of Hwy 7 and the river. He then informed him that all the bombing targets had been prepared by the Koreans Marines, and to make them more realistic they had placed captured explosives on them so the pilots would see secondary explosions.

When Gilbert called the spotter plane he broadcast so that the fighter bomber pilots could hear his instructions. He told them that the Korean Marine Division had placed targets with secondary explosives for realism when they were to hit. Do not fire on soldiers on the ground as they were Koreans on maneuvers. This is a training exercise under control of the Koreans but beware of returning fire from the secondary explosions. Do not log combat time or give body count estimates.

The Colonel listened to Gilbert then held his finger up and circled it around his ear. The battle was on, and the Colonel was busy. Where Gilbert could see the ground it was smoking from the culverts full of fu gas that had been dropped from an Army Chinook. As the ground was burning the Cobra gun ships came in and tore the earth up. It started to look like a freshly plowed field with outcroppings of rock and small islands of trees or brush. The enemy had disappeared under ground in their tunnels. Slick after slick (UH-1 helicopters) came in and lifted right off to get out of the way of another UH-1. This continued until the entire Marine Division was on the ground.

While this was going on Gilbert answered a radio call form the USAF Base Commander. "I just got a phone call from Saigon Special Operations. They said that there were no combat operations authorized in this area and want to know what is going on." Gilbert answered, "Colonel, I don't know of any combat missions. I'm in the Korean Marine Division Commander's CC Helicopter, and have been giving the USAF training flights the targets he has pointed out. He sure has gone out of his way to make them realistic but it's still only training. What are you going to do?" "I've already ordered all squadrons to scramble training flights, and I'm talking to you from one of my squadron's planes. This is a chance for me to get combat currant too. There will be ten flights of four planes on station with in ten minutes, out."

As he looked down every one of the Korean Marines had disappeared except for two or three standing around holes or openings in the ground. They had all gone underground after the North Vietnamese solders. Gilbert glanced up at the S-2 Officer and somehow he looked different. The Captain was looking at him and grinning from ear to ear. Gilbert started to look away, and then he realized the Captain had on a Major's leaf. Gilbert then looked over at the Colonel to see if he had noticed what his S-2 Officer had done, and he was wearing a new Star on his uniform. It appeared that somehow they had been promoted while he was busy convincing everyone that this was just training for his class of Cambodian soldiers.

As he watched the Korean Marines started coming out of the holes dragging bodies after them. Their helicopter landed so their new General could greet his victorious marines and the General said. "We have found the enemies headquarters. You check each body to be sure we have found the spy you are worried about. When you find him I'll put them all back in the ground and seal the entranceways. If you ever get to Korea I'll give you a medal for the best training mission we have ever been on." Gilbert looked at thirty bodies before he found the informer. Now he was satisfied that he had done all he could do to save the lives of his PRU Tiger Scouts and their families.

While the marines were putting the bodies back in the holes and blowing up the entrances to seal them he called Bill Colby from the helicopter. He told him that he was with the new Commanding General of the Korean Marine Whitehorse Division on a training exercise, and that the USAF was helping by conducting new pilot checkouts on practice bombing missions of special prepared targets. Colby asked, "Is our VC Informer involved?" Gilbert answered, "He played a major part. The last time I saw him I became convinced that most if not all of my PRU Tiger Scouts and their families will survive. I'll phone you tomorrow with more details."

The helicopters were now picking up the Marines and returning them to their compound. As they took off Gilbert asked the new Major S-2 over the intercom, "Where are the medi-vac helicopters?" The new Major answered saying,

"We had a few men wounded but they can still fight and are returning with their units." Gilbert then asked, "Any prisoners?" The General came on the intercom and said, "This was a training exercise for your students so how could there be prisoners?"

He was right of course. The Navy wasn't concerned as their mission included security on all waterways and support of the Province Recon Units (PRU) Tiger Scouts. The only thing different about today was that their instructions were to turn all prisoners over to the Korean Marines instead of to the Province Interrogation Center (PIC). The same thing applied to the Army's Military Police. They controlled the traffic on Hwy 1 and Hwy 7, and any Viet Cong detained had been turned over to the Korean Marines. The Marines interrogated them and disposed of them in their own way.

As Gilbert walked out to his pickup truck the Marine S-2 Officer went with him, and Gilbert speaking in German asked about his promotion. He answered, "The General was told two years ago that when he found and destroyed the North Vietnamese Camp he would have his star." The General had also told him that his promotion would only come when he found the enemy Camp. He had been carrying his oak leaf insignia in his pocket for over a year, and when he saw the General putting on his star he had put on his Major's insignia. Gilbert thought to himself that this was fair.

This had been a long day for Gilbert. Actually it had been a long week, but with the help of Bill Colby, the USAF, the US Army, and the Korean Marines he had saved the lives of his PRU Tiger Scouts and their families. Best of all there was no longer a Viet Cong or Communist presence in Tuy Hoa. In three hours a North Vietnamese HQ & HQ Battalion, two ammunition dumps, a fuel storage dump, the Viet Cong Communist Headquarters for the Sapper (terrorist) and recruiting had been destroyed. It was too bad that the Americans at home would never learn the truth about how the Communists were beaten. It was a battle with a tremendous victory for the Korean Marines that will go into their history books, but not in America's history as it was only a training exercise.

The USAF records clearly show that training flights for new pilots took place at Tuy Hoa, and Navy logs record that the Swift Boat patrols that day assisted a training exercise and maneuvers of the Korean Marine Whitehorse Division. Bill Colby was one of the few people who knew that the North Vietnamese Army had a double agent being paid by the Province Officer In Charge (POIC) of Tuy Hoa. He maneuvered events where Gilbert would react before any spies in MACV or with Special Ops in Saigon could warn the Communist of their danger.

The following day Gilbert phoned Colby from his Province Interrogation Center (PIC) office, and told him about how the USAF had supported a Korean Marine training exercise. Also how they had helped save the Koreans from any losses. Bill answered saying, "This will have to stay as a training exercise. No medals or combat time for anyone. It never happened. Let the North Vietnamese worry about what occurred." Then he asked, "You said that there were no prisoners taken?" How many do you estimate got away?" Gilbert said, "I'm hoping that the couriers made it. We gave them three hours to get away, and if they made it there is no way any of them can know what really happened. The USAF destroyed at least three major supply dumps and bombed the jungle west of the North Vietnamese Camp forcing everyone fleeing back into the killing zone. The Navy did the same thing from the river, and the Army took care of Hwy 1 and Hwy 7. It looks like a total wipe out."

"Finish your training of the Cambodian Team and keep me posted. I need you in Nha Trang as soon as possible," Colby said. "What do you want me to do when I get there?" Gilbert asked. "You are to set up the same program as in Tuy Hoa," Bill replied. There would be no medals or recognition of what Gilbert had accomplished, and the records would only show that it was a training exercise but on his next pay accounting report there appeared a notation that he was now a GS-13 which meant a pay raise.

After the phone call to Colby he drove to the Airbase arriving about 1000 hours. The Cambodian soldiers had just started to work on their weapons when Gilbert arrived. He asked Lieutenant Youeca if he thought that they were ready and upon his affirmative answer said, "Men, prepare for inspection of all your equipment." When they were ready he checked every weapon and item of equipment they would take with them on their snatch and grab raids in Cambodia. Gilbert then tested each Cambodian soldier on his understanding of the equipment and the mission. By 1400 hours he was satisfied, and while the Lieutenant gave them a chance to play volleyball Gilbert walked over to the Officers' Club. Then he phoned Colby, "They are ready." Colby answered telling Gilbert that there would be a DC-3 plane arriving at 1000 hours the next morning with another group of Cambodians for training and to pick these men up.

Colby also informed Gilbert that the Lieutenant would remain at the Tuy Hoa Airbase to help train the Cambodian soldiers and the men he had trained would report to the Cambodian General Um Savuth at Penh City, Cambodia. Gilbert's contact would be a Special Forces Master Sergeant William Waugh, the Cambodian Division Advisor. Gilbert was sure now that Colby had some plan so that the world historians would someday learn that the Dark Force and their Communists allies could be defeated. Many of Gilbert's informers who worked just for the money were quitting. All gave the same reason saying that the Communists were winning the war, and the Americans were giving up. Only the school teachers and professional people never gave up but they were worried. Gilbert knew that if the rumors were true, and the Communists took over they would all die unless they could escape to the United States. The Paris Peace Talks had turned into the Paris Peace Accord which was another name for a document giving up the rights of all Christians in Vietnam.

The patriots in the CIA were banned by law from helping the American people within the United States. They knew the danger of the Communist attacks, and understood how America was the real battle ground between the Dark Force and the Bright Force but were powerless to intervene.

In 1967 Gentle Ben had became concerned about the targets that the B-52s were bombing in North Vietnam. Gentle Ben and Helen who were in Bangkok, Thailand at the time decided to visit Da Nang, Vietnam. It would give them a chance to surprise their son Ben (Bud) Jr, a Navy CPO stationed at the US Navy Camp Tien Sha at Da Nang. Upon returning to Bangkok he wrote two papers on the Vietnam War. At the time Gentle Ben was flying as USAF Major checking out Thai Marine Police pilots in their L-4 Lake flying boats. One paper titled "Shangri-la Vietnam" was printed in news papers around the world, and the other paper titled "Wrong Target" was classified secret.

Upon returning to the United States Gentle Ben phoned Harry Alexandra, his old OSS mentor, at his home in Ocean Springs, Mississippi telling him about the thesis titled "Wrong Target" that he had written in Thailand. It was about his research on bombing targets in Hanoi, Vietnam. Gentle Ben, posing as a newspaper reporter with a letter from the South Bend Tribune, Indiana had joined a chartered Japanese Airline plane for a flight with other Japanese reporters to view a B-52 bombing mission over Hanoi.

After the bombs were dropped the Hanoi Airport Tower cleared the Japanese plane to circle at ten thousand feet to take pictures. Gentle Ben wrote that right after the bombs hit and the dust settled the craters looked as if they had been there for days. Now he wanted to observe actual targets before a bombing mission was sent to destroy them. Two days later he was escorted to a factory making anti-aircraft weapons. He found that the employees, their families, and their children all ate, attended school, and lived in the factory. Only the Communist bosses, factory supervisors, and government people lived in subdivisions or homes away from the target area. The thesis titled "Wrong Target" documented that the targets should be changed to the housing areas. Kill the Communists and let the people alone or something like that.

Harry asked, "What did you do with the story and thesis after you wrote them?" Gentle Ben answered, "I put them in the DROP that was located in the Bangkok National Airport." "I would have loved to be there when the CIA Agent making the pickup read your papers. Did you sign them?" Harry asked. "Only the story. On the thesis I just put my control number and addressed it to President Johnson." Gentle Ben answered. Harry started to laugh and then he asked, "Hasn't Helen had lunch with Lady Bird in the Rose Garden? Then he said, "If I remember correctly you even flew a helicopter carrying a news man covering President Johnson a few months ago." "Yes, that's true," Gentle Ben answered.

"I've read your story in the newspapers. I doubt if the President ever saw your thesis but even if he did he wouldn't like someone telling him he had picked the wrong targets. Did you send a report to the South Bend Tribune?" "No, I didn't. I just billed the Japanese charter plane to the American Embassy." Gentle Ben replied. "I phoned you because I just got two letters from the Air Force, one promoting me to LT/Colonel, and the other saying that I'm being retired with twenty-two years service. What do you suggest?" "Give me a few days and I'll get back to you. I'd start thinking about Saudi Arabia or Vietnam with one of the Company's contractors," Harry laughed as he hung up.

Now here he was in Vietnam flying as a CW-3 for the 1st Cavalry Division supporting Gilbert a man who never knew when he was well off and sure didn't know when to quit. Here was a man who had just help defeat a North Vietnamese Regular Army Battalion and had destroyed the Communist Viet Cong organization in Tuy Hoa all while in a cast from being injured in an ambush by the enemy. Now he was training Cambodian soldiers to start making the Ho Chi Minh Trail impossible for the North Vietnamese to use.

The fortunes of war are unpredictable at the best of times. Two weeks after the first Cambodian team had left on their mission Special Forces M/Sgt Waugh contacted Gilbert telling him that the team had all been killed except for one man. This soldier was being sent back to Gilbert to join in the new team he was getting ready. After they had met Gilbert's tests and were ready the soldiers were sent to report to General Um Savuth in Penh City. Once again Gilbert received a message from the Special Forces Advisor M/Sgt Waugh that the team had been ambushed only this time with no survivors.

Gilbert went over to the Officers' Club and phoned Colby telling him about the message. "How is your back?" Colby asked. "It not bad." Gilbert answered. "Good, there will be a DC-3 arriving in two hours to pick up the class and the Cambodian Lieutenant. They will be traveling Black and there is no need for you to know their destination," Colby said. Then using Gilbert's given name continued, "Joe, we think the North Vietnamese may be using you as a Tracker. With the injuries from the last time you were ambushed they may believe you are out of the action, and by watching you they can target the teams you are training. When you are ready notify the Province Officer In Charge (POIC) that you are going on R&R. Just make it your new assignment location. Good Hunting."

It was a shock for Gilbert to learn that he had become a Tracker for the VC yet even he could understand that the Communists were confused. The Communists do not have individuals who make decisions with out direct supervisions of the Peoples Revolutionary Party (PRP) Committee or at least a Senior Communist Party Member. They had no comprehension of anyone working without supervision or specific orders. In their society these people were killed, for if they were allowed to live they would surely try to assassinate the party bosses to take over their positions. The patriots in the CIA understood this and use it to the United States advantage. When they found such a man or woman they watched over them and used their talents.

Sometimes even the person never knew. Other times they would offer the man or woman the position of a contract employee or possibly the person would be employed by a company holding contacts with the government. The individual may already be in the military service of the United States. A few others would receive mentors who were senior members of the CIA. Nothing in writing, only verbal suggestions, and favors. Gilbert was one of these men who after twenty years in the USAF became a CIA contract employee assigned to work with the Department of the US Army in Vietnam.

When Gilbert woke up it was 0800 hours and late for him. This was his last day in the French Mansion, and he strolled into the kitchen looking for something to eat. The old Chinese housekeeper and cook made him a full English breakfast and fussed over him. She seemed to know that it was the last meal she would prepare for him. After she asked the second time what he wanted for dinner that evening he finely said, "I'll be leaving for Saigon for a few days of R&R today. Don't worry about my meals until I return."

He packed a bag with two changes of clothes and his toilet things. Next he put two letters from his family at home on his table in the bedroom, and then left a half finished letter he had been writing beside them. The only shoes he took were on his feet. His combat boots were left under his chair, and the flak vest hung over its back were he usually kept it. An AK-47 that he had picked up in his last firefight was in a corner of the room. He stopped and looked the room over one more time to be sure it looked as if he had only left for a few days. Then he reached into his pocket and took about twenty dollars in Dong (Vietnamese money) in small bills and folding them over put these under the corner of a half empty scotch whiskey bottle on an end table. Now it really looked like he was only going to be gone a day or two.

Next he took the PRU pickup and drove over to the National Police Compound where he used the phone in his office to call Air American operations asking for a flight for the "Hotel Man" with pickup code #103. They must have had a plane making a drop off in the nearby area as they said the plane would arrive in thirty minutes. He told his secretary that he was going on R&R to Saigon, and to take some time off. She answered informing him that she, her husband, and her two children were leaving to visit relatives in the United State and wouldn't be here when he returned. While they were talking the Special Force PRU Advisor had seen his truck and stopped in to say hello. Gilbert asked him to drive his truck and take him to the airstrip to meet the Air America plane. As they left the Police Compound the Sergeant asked if he needed to sign out with the new Province Officer In Charge (POIC) before leaving. Gilbert told him that it wasn't necessary as he wasn't going to be gone long.

As they passed over the rail road track and stopped at Hwy 1 waiting for traffic to ease up Gilbert couldn't help but look for a mark on the electric pole to see if anyone had left a message. It seemed like only yesterday when a mark on that pole had changed his life forever. He was no longer just a CIA Interrogator but a man who had accepted the responsibility of the people assigned to him. He had measured up and made the hard decisions that only a leader must make. As they drove across the road and turned to park beside the airstrip the Air America plane turned on a short base leg. The pilot never shut the engine down as Gilbert threw his bag in, slid into the seat, and fastened his safety belt. The pilot didn't even taxi but advanced the throttle, and took off from where Gilbert had got into the plane. The plane climbed and turned east toward the South China Sea. The pilot said, "We'll stay over water so we won't have to worry about ground fire." Gilbert laughed and said, "Don't worry there are no Viet Cong around here."

As they turned south Gilbert told the pilot to land him at the airport in Nha Trang. As they were taxing in Gilbert asked the pilot keep his name on the manifest to the Tan Son Nhut airport and not log the landing in Nha Trang. The Air America pilot was used to strange requests from his passengers (spooks he called them) and just nodded. If the Viet Cong were using him as a Tracker then let them chew on this. It must have worked because it took Colby three weeks to find him, and he knew where to look. This was a lot more dangerous place than Tuy Hoa with a lot more Viet Cong or maybe he had just forgotten the ambushes he had lived through during his last assignment.

He waited until the Air American plane took off before he picked up his bag and started walking to the guard post at the entrance of the airfield. Gilbert had removed his shoulder holster placing it in to his bag but kept his 25 Cal. Browning automatic pistol in his left pocket. It was small enough that he could stand with his left hand on the gun and while waving his hat with his right hand, fire the weapon hitting anything in a room. He had ruined a lot of pants in this practice, and he was now a very deadly shot.

Gilbert stood at the gatehouse of the airport waiting to hitch a ride on any Army truck that was heading into the city. He had asked the White Mice (Vietnamese Police) who were checking the trucks where the Officers Special Assistants to the US Ambassador (OSA) Compound was located. The guards said that they would ask the driver of the next truck that would be passing the OSA compound to give him a ride. The very next truck approaching the gate was an US Army 2 ½ Ton Truck half loaded with canvas tents and cots. The driver was an ARVAN Army Private with an ARVAN Guard carrying an M-16. One of the gate guards told Gilbert they would drop him off at the OSA Compound.

This was lot different than his last assignments were he drove his own vehicle and someone would take him around to help sign him in while introducing him to everyone. Now he was trying to move without the VC being able to track him. Before it was only ambushes he had to contend with but now Gilbert didn't want to expose the people he was working with. It was harder but Gilbert liked the taste of the excitement that comes with blending in and disappearing into the background. As the truck stopped at the OSA Compound gate one of the White Mice on duty started asking the driver about the tents while another just glanced at Gilberts CIA ID walking on water card and waved him through. No one paid him any attention as he walked, carrying his bag in his right hand, and with his left hand in his pants pocket to the Province Officer In Charge (POIC) Office. The secretary asked him to sit and wait while she informed the POIC that he had a visitor. Gilbert was sitting with his bag at his feet with his left hand still in his pocket when the POIC walked out and realized who he was.

The man greeted Gilbert and invited him into his office. He was very professional and said, "I've been expecting you. I have read some of the reports of your work in Tuy Hoa. Mr. McAffe, the Ambassador's special Assistant, has asked me to give you these special account numbers for drawing funds for your projects." Then he handed him a card with the two numbers. The one number was for drawing Embassy funds, and the other was for Army Finance Center draws. Then he said, "I understand that you are going to help our Province Interrogation Center (PIC) become more affective. It's great to have you here and anything I can do to help just let me know." Then he asked, "Saigon Special Operations said you were on R&R somewhere in Saigon. How did you just appear waiting in my office?" Gilbert laughed, and told him that he was practicing some new techniques which he hoped wouldn't bother him too much. Then Gilbert asked if he could be allowed to use a safe house reserved for visitors and not be noticed. The POIC was an old experienced CIA man, and he just nodded and asked? "Anything else?"

Gilbert answered, "I'll need a jeep and a good interpreter proverbially one with combat experience." The Province Officer In Charge (POIC) nodded and said, "I have just the man his name is Hung, and he is a little shorter than you are but a little cocky." When Gilbert first met him the man was carrying a Colt Automatic Rifle (Carbine 15) with a thirty round clip and a folding stock. Hung looked and acted like he enjoyed using it. To get better acquainted with his new interpreter he invited him to pick the best restaurant in town and have dinner with him. The man hesitated but when Gilbert added I'm paying for it. Without further hesitation Hung answered, "I thought you would never ask!" Gilbert thought to himself that this man had been around Americans too long but he liked him.

Hung took him to the best French restaurant in Nha Trang. It catered to Americans and the rich, very rich Vietnamese. When they entered Gilbert asked the waiter to seat them toward the back where he could sit with his back to a side wall. Gilbert was still trying to fade into the background but it was hard to do with Hung carrying his mean looking CAR-15 in one hand telling people to get out of his way as he lead them to the table Gilbert had picked. It was also a very good thing that he had been given his own funds accounts as the meal of one lobster for himself and a steak for Hung was going to be a months pay. The lobster overhung the platter on both ends, and the steak would have made a Texan proud. This was a seven course meal, and Gilbert noticed only one item that had a cost listed on the menu. It was for a bowl of soup, with the cheapest selection costing $500 Dong, and the costliest over $1000.

This wasn't Saigon but with the prices on the black market controlling the economy, inflation was ever 100% per month. It didn't matter what food costs in Nha Trang or Saigon Street orphans are always hungry. Gilbert and Hung had eaten their full and were just sitting back to enjoy drinking a beer when Gilbert noticed little hands coming out from under the table reaching up trying to get scrapes from his plate. One of the waiters came rushing over hollering for the children to get out. Gilbert waved the waiter to stop but it was Hung's CAR-15 that caused the sudden disinterest in their table by the waiter. Gilbert pulled up the table cloth and looked under it. There were three children with dirty faces about five or six years old with bread clutched in one hand, and the other stuffing scraps of food into their mouths.

Gilbert had put his back against the wall about two feet from the table and Hung scooted his chair so his back was against the same wall. As they drank their beer watching the little arms and hands dart out and fish for food Hung stared cutting meat from the bone of his steak leaving it scattered on the plate. Gilbert asked, "When I looked I only saw three children but there are enough arms and hands for five. How many do you think there are?" Then he said, "I wonder how they are getting under the table?" Hung pointed at the door and said watch, "Here comes one now."

A child of about four had crawled in behind the guard or waiter at the door and was moving fast along the wall. Another child was crawling just as fast along the wall in the opposite direction leaving the restaurant. They were like ants taking turns at the table. Gilbert and Hung just sat drinking beer and visiting. As people started leaving the restaurant they all stopped, bowed, and laid some food on their table. A part of a dinner roll or a piece of meat but it was always something that the children could eat. Hung never took his eyes off the waiters or the door, and he never let his CAR-15 drop. When the little hands stopped reaching for food and the dinner guests had all left Gilbert called for the manager. When he arrived Gilbert asked for the addition (check). The manager said, "The weapon is not necessary. Some Americans do not like to see orphans beg while they are eating. We always give the children any food that is left. Will you pay for the beer? There is no charge for the food. You have been our guests." He bowed in the Buddhist way, and as they left all the waiters bowed.

Gilbert had learned a long time ago that one should not look a gift horse in the mouth. In other words why would someone want to know the age of the horse if it was a gift or something like that? It had been a long time since Gilbert had been in Saigon and observed the Dark Forces at work. It only seemed like yesterday when the street orphans had been slaughtered by a Viet Cong mortar attack in Saigon, and he had promised God that night he would help the street orphans. This wasn't much but it still made him feel that his work was worthwhile.

The next day he went to the National Police Station to meet the Police Chief and to get acquainted with their methods of operation. He was surprised but the Chief of Police and all the White Mice seemed glad to see him. They greeted him warmly and were trying to please him. He was wearing his shoulder hostler over his white shirt with a Belgium 7mm automatic pistol so all could see it. The Browning automatic was still out of sight in his pocket. Gilbert was so surprised by the attitude of the police that he looked back at Hung who was following him to see if he was doing something to cause this reaction. The man looked as surprised as Gilbert and had his CAR-15 with the stock folded hanging over his right shoulder.

The Police Chief greeted him saying, "You are a brave and kind man. Going to a French Restaurant with no weapon and spending all that money to feed the children has gotten everyone's attention." This was not really what Gilbert had planned. He had really planned to stay in the background. He answered saying, "Hung was there with his CAR-15 so I was in no danger. The children were hungry." The Police Chief looked at Hung and said, "He can shoot Viet Cong with that thing but someone has got to point them out." It seems that Hung had a reputation. Then he asked Gilbert what he needed. Gilbert answered asking, "Can we speak in private?"

After they were alone the Chief of Police turned on a squeaking fan and then speaking softly again asked how he could help. Gilbert suggested that he should plan for an operation to round up some VC suspects in some area that was known as Viet Cong territory. The most important thing was that only the Chief of Police would know the target. Only Gilbert and the Chief would know the day and time when the police would strike. Gilbert would provide the funds for the operation, and there would only be twelve policemen besides the Chief, Gilbert, his interpreter and bodyguard Mr. Hung. It was agreed that Mr. Hung would drive the police truck and the policemen would be selected as they went out and loaded into the truck. The Police Chief would ride in front with the driver and Gilbert would ride in back with the policemen. The Chief would direct the driver to the target and that way no one would know in advance where they were headed.

It was to be a quick snatch and grab raid to obtain prisoners for Gilbert to interrogate. Then he could start planning how to eliminate the Viet Cong presence in the city. The Chief of Police knew of one man in a small hamlet of about eight houses that was an active Viet Cong member. The Chief knew the man's name and had a picture of him shooting a policeman who was bound hand and foot. They had tried to capture him before but the man seemed to know when the police were coming and disappear. The next afternoon Gilbert stopped into the station just as a change of shifts was going on. The Chief called out to Gilbert, "I've got a new truck. Would you like to see it?" It was the signal they had agreed upon. Gilbert was to walk through the men going off duty and pick twelve policemen to join the Chief waiting by the truck.

As Gilbert followed the last man out to the truck the Chief of Police was grinning as he made sure each man had picked up a M-16 with two clips from the Sergeant in charge of their armory. It was beautiful plan. Only the Chief of Police knew the target area where the VC might be found. Not even Hung who was driving knew where they were headed yet they all knew it was into a firefight with the Viet Cong.

CHAPTER SIX

It was late in the afternoon during the heat of the day when a one ton Ford truck normally used to carry rice straw was seen speeding down the main road of Nha Trang. The bed of the truck with its four foot high sides was filled with a National Police Snatch & Grab Team who were on their first Dai Phong (Big Wind) Mission. Combat Interpreter Hung was driving, and the Chief of Police was in the truck's cab giving him directions. Gilbert was standing in the back with the other policemen hanging on as Hung swerved around and through the traffic.

Gilbert had deliberately picked policemen of about his height, and had everyone wear the same uniforms that were called tiger stripes. His uniform was almost identical as the police only his had been sanitized (no markings, name, or identification of any kind). The weapon he carried was an M-16 with a thirty round clip and a small 25 caliber automatic pistol in his pants pocket. All the policemen were armed with the US Army M-16 with the 15 round clip. His plan was to blend in with those around him and not to become a target or stand out, so the Viet Cong could track him.

Colby had told him that he may have become a Tracker for the Viet Cong. Also that the VC were watching him and making a list of which Vietnamese were helping the Americans. The Paris Peace Talks was becoming the Paris Peace Accord, and the North Vietnamese were planning ahead to the day when the United States would abandon Vietnam. The Communists have long memories and have a final solution. It is a simple plan, and the only ones who could stop them were the US Military fighting men and women. The Viet Cong would first kill all those Vietnamese who were Christians, and then continue until everyone who had helped the Americans were dead. There would be no one left to rise up against them, no underground, no rebels, and the Dark Force would then rule the Garden of Eden.

The truck proceeded about two miles north of the city to a Bailey bridge that had been blown up by a Viet Cong sapper (terrorist) squad, and then it turned east on a dirt road leading to a small fishing hamlet of eight houses. The Viet Cong were caught completely by surprise, and no one had a chance to escape. The police lined up everyone on the road, and the Chief of Police personally looked into the face of each man while studying the photograph of the man they were looking for. There were three men and eight women with a bunch of small children standing on the road. Gilbert was close to the Police Chief and Hung stayed back a few feet watching and guarding Gilbert but close enough that he could interpret what was said if needed.

The Viet Cong they were searching for was not amongst the men standing on the road with their families. The Police Chief must have had an informer who had given him the tip that the VC they were looking for was in the hamlet. He sure wasn't happy when he ordered six policemen to start searching the houses. The homes in the hamlet were not large, most consisting of three or four rooms, and thatched roofs of palm leaves except for two that had metal roofs. As the houses were being searched Gilbert asked the Chief of Police, "Do you trust your information?" The Chief answered, "He's here." Then Gilbert suggested, "Why don't you put three policemen at each end of the road where they can see him when he runs out of his hiding place?" and then he added, "Set fire to all the houses and wait." Now Gilbert was an interrogator first, last, and always. When he made the suggestion to the Chief of Police he had moved to where he could watch the men being held. He knew that the Viet Cong women were the real heads of the households and would never blink an eye even if the police burned all of the houses looking for the man.

Gilbert knew that somehow the women would make sure it was the men's fault, and they would feel the pain of losing their homes. Sure enough one of the men pointed to one of the larger houses with a metal roof using a finger while scratching his crotch. He didn't move his hand or even look at the house just quickly pointed and went back to scratching himself.

The Chief of Police was ordering torches to be made ready when Gilbert asked, "Let me take six policemen and search that house again, and (pointing at the house). Maybe I can get lucky?" Upon getting the go ahead from the Chief he had the six policemen find seven slim pieces of bamboo and then sharpened the ends to a needle point. When they went into the house he had all the doors and window opened so they could have as much light from the afternoon sun as possible. Then they stuck the rods between the floor boards and down into the earth with no luck. The policemen really got into it as they understood that it was behind or in something that the man must be hiding. They searched in, behind and under everything and still couldn't find him.

They did discover a dried up dead rat but no VC. Then Gilbert looked up and right above their heads were the roof rafters and lying across three of them in the center room was a rolled up reed rug. One of the policemen poked the rug with his stick and it moved. All the other policemen rushed over and pushed at the rug with their sticks until it fell between the rafters to the floor. They had found their Viet Cong.

The policemen started hitting the rug with their sticks until Gilbert straddled the rolled up rug hollering for them to stop. With all the noise the Police Chief and Hung came rushing in to help. All the policemen were patting themselves on the back and everyone was laughing. They wanted to take and tie the prisoner in the back of the truck where everyone could see him as they drove back to their compound. Gilbert final convinced the Chief of Police that it would be dark soon and it would only alert the Viet Cong that they had an informer in their ranks. Besides it would be night before they got back to the police compound, and that would surly give the Viet Cong time enough to set up an ambush if they knew the man was their prisoner. Gilbert suggested that if the man was rolled up in the rug and laid in the back of the truck they might be able to get him to the Province Interrogation Center with out a firefight.

It seemed to be a hard decision for the Chief of Police to make. They had finely caught the man who had his picture taken as he murdered a police officer, and here was Gilbert wanting the police to protect the prisoner. The Chief was an intelligent man, and he suddenly realized that the light was fading. He just nodded and started barking orders.

They wrapped the prisoner back into the reed rug, tied it at each end, in the middle, and then threw it into the back of the truck.

Now Gilbert had to convince them not to burn the hamlet. Light was fading fast, the Dark Force would soon be in control, and here they were in Viet Cong territory arguing about burning some houses. The Chief of Police spoke good English and Gilbert told him that his informer was the one who had signaled him which house the VC was in. Then Gilbert told him that he had promised that they wouldn't burn the houses if he was found. The Chief just looked at him wondering when Gilbert had talked to his informer but when Gilbert reminded him that in another few minutes they would have to fight their way to the National Police Compound they all climbed into the truck, and almost before Gilbert got aboard Hung was driving wildly back to the city.

It was a highly successful mission and not a shot had been fired. As they drove into the police compound Gilbert expected them to stop at the Province Interrogation Center (PIC) but instead they parked at the police building next to the jail. Inside the National Police headquarters' building they had a holding cell large enough for a dozen prisoners. It was in the center of the room made of closely spaced steel bars from the concrete floor to a ceiling of wire mash below the room lights. There were no cots or latrine and the police desks with work spaces were located around the holding cage. The cage had fifteen prisoners in it when they arrived, and with the Chief giving orders its door was flung open. All the prisoners were actually thrown out of the building. Gilbert was completely ignored, and Hung whispered, "The Chief is very pleased but maybe we should leave. I think that this is the police's get even time." Gilbert nodded as they moved away from the excitement.

"Let's go over to the Province Interrogation Center (PIC). Maybe someone is there, and I can get a feel of where we will be working," Gilbert said. "I don't think that this VC will get that far," his Interpreter Hung answered. As they walked toward the building Gilbert could see that it was drab looking building with low ceilings and made of concrete. Twenty feet away the smell was so bad one could almost see the fumes rising from the building and it stunk of death. As they walked into the building they saw that the inside was nothing but a damp, dirty, hole that stunk of blood, puke, and shit. There was no one in the building so they left, and turned back toward the police building.

"Hung, go get the jeep and pick me up while I wait here. We will go back to the safe house and comeback tomorrow when things have settled down." As they drove toward the compound gate they had to pass the prisoners who were now sitting in the driveway. There were no guards but Hung just pointed toward the towers. The White Mice (police) in them were all watching the prisoners, and as the jeep passed they could hear God awful screams coming from within the police headquarters' building. Without asking Hung said, "If they stay without calling attention to themselves the police will open the compound gate and drive them out tomorrow morning. If the VC doesn't last to morning they may get to join him."

Gilbert spent the next day scouting the city and meeting the Special Forces Advisor of the local Province Recon Unit (PRU Tiger Scouts). This was the routine he followed in every city that he worked in. He thought about phoning Colby but there was nothing to report and no reason to alert the enemy spies in Saigon that he was here. He was sure that the snatch and grab raid he was on would never be connected to him as it had turned out to be a typical Vietnamese police action. It was the second day after the capture of the VC before Gilbert returned to the police compound. Keep his head down and blend in, Colby had advised, so he did.

It was 1000 hours on a beautiful sunny morning two days later when Gilbert and his Interpreter returned to the National Police Compound. The first thing he noticed was that there were no prisoners in the court yard or on the compound street. They were gone as Hung had predicted. As they walked into the police building the second thing they noticed was that there was only one prisoner in the lock up or holding cell. It was the Viet Cong that had been captured.

He was naked and handcuffed with his arms held to the bars of the cell like Jesus on the cross but his feet were on the floor. There was a bucket on the outside of the cell about three feet from the hanging VC that the policemen pissed into. There was also an electric welder in the room with its ground connected to the cell bars. The prisoner hung there unconscious with the imprint of cell bars on his front and back. The imprints appeared to have had been put there with a branding iron but there were no such tool in the room. Gilbert had no idea of what the police had used the welder for or how the prisoner had gotten the burns.

Gilbert asked to see the Chief of Police and was immediately taken to his office. After the greetings were finished the Chief said, "You should have visited yesterday. The VC was in good voice, and he changed his religion three times." Then he asked, "What can I do for you today?" Gilbert then gave him money as payment for the capture of a Viet Cong, and then told him that he would pay for the clean up of the Province Interrogation Center (PIC). He stressed that he would pay to have the PIC painted white and for new white coats for the interrogators. Gilbert also asked that the outside of the building be painted white and the grounds around it cleaned up.

The Chief asked, "Wouldn't it be better for the interrogators to wear red jackets to help cover the blood?" Gilbert then explained that there would be no blood as the new system of interrogation he would teach them was without torture. It was more like what the police used in the United States. The Chief of Police seemed unconvinced until Gilbert told him that he would pay for everything. Then the Chief told him that the work would be completed in a few days, and Gilbert could pay him when it was completed.

As they were talking suddenly there was a scream from the prisoner holding area. The Chief grinned and said, "You are in luck. Our cop killer has at least one more chance to experience hellfire before he gets there. Do you want to go and watch?" About that time the smell of hot urine filled the air. It was like the smell of someone pissing on the hot coals of a fire. The police had turned on the electric welder so the bars of the cell between the positive and negative connections were turning red hot. Then some one threw the bucket of piss on the prisoner. Just before Gilbert left he told the Chief of Police that he would pay him for every VC captured and delivered to the Province Interrogation Center (PIC). If any of the prisoners were tortured before they were delivered to the PIC there would be no money. He also wrote in his Journal that night; I pray that the Lord will forgive me for not letting the police kill the VC when he was captured. Somehow he had missed the signs that this was a personal vendetta between the police and this man.

Two weeks had passed and the Province Interrogation Center (PIC) was completed and ready for work. Gilbert had rooms built to hold prisoners in the Province Interrogation Center (PIC) where they could sleep and be fed without leaving the building. The idea was to give them a sense of security if they turned and started to help the police. The prison was in the same compound area, and Gilbert wanted to keep those that he could use from contact with any other prisoners until after they were convicted and sentenced by the Vietnamese Court. If they became members of the Province Recon Unit (PRU) Tiger Scouts or informers the Vietnamese Court would release them from custody and sometimes grant pardons. Many informers never entered the court system but worked under the protection of the Special National Police.

Gilbert was teaching a class to the Province Interrogation Center (PIC) Interrogators on how to interrogate a prisoner with out using impact torture methods when the Chef of Police walked in and asked, "Would you like to go on another snatch and grab raid?" Gilbert stopped teaching and said, "Yes" but then added, "Let's talk in your office with the fan on." The Chief laughed and instead led him to the center of the parking area. "This is better.

No noisy fan and no one can hear us. Do you pay more for big fish?" He asked. To an interrogator this was just the opening of negations. Gilbert hesitated a moment and then said, "I think something can be arranged." He wrote in his Journal later that he knew this had to be a really big fish. Maybe even a Communist Peoples Revolutionary Party (PRP) member. If it was he had to find some way to interrogate the Viet Cong before the prisoner could be transported to the Saigon National Interrogation Center (NIC) which still used the North Vietnamese impact method of interrogation, and the black box interrogation techniques of the career (Jewelers) CIA Agents.

The next day the Police Chief and Gilbert started planning for the raid into Viet Cong territory. Gilbert learned that the tip had come from the same informer that had led to the arrest of the man who had killed the police officer. The man had said it was a thank you for not burning his house down. This raid was going to be different in that the police would not be involved. The Chief gave Gilbert the location where the Vict Cong was going to be but didn't have a time for when the target (big fish) would be in the area. It would be Gilbert's party. He just knew it was a big fish. It just had to be if the Police Chief wanted to be close enough for credit and far enough away not to be involved if things went wrong.

Gilbert contacted the Province Recon Unit (PRU) Advisor, an Army Special Forces Sergeant and told him that he would need enough PRU Tiger Scouts to capture a high ranking Viet Cong plus his bodyguard of at least four men. They decided that if the Special Forces Sergeant took two PRU Tiger Scouts and engaged the bodyguards after they had passed a bend in the road leading toward the fishing hamlet that the other guards would surely go into a defensive posture. This would force their target to retreat down a side trail leading towards the water. Gilbert would be waiting in ambush with the ARVIN PRU Chief and six PRU Tiger Scouts. The plan was for them to snatch and grab the Viet Cong before falling back to their truck. The Special Forces Sergeant was only going to feint his attack and not actually engage the enemy. After grabbing the Viet Cong Gilbert was to pick up the Special Forces Sergeant and the two PRU Tiger Scouts as they passed the bend in the road on there way back to the city. The Gods of war must have laughed as they watched the feeble attempt of Gilbert and the Special Forces Sergeant trying to guess what the Gods of War were going to do in the firefight that was coming.

The next day the Chief of Police stopped Gilbert as he was walking to his training class in the Province Interrogation Center (PIC) building and told him that the target was going to be in the area at 1600 hours tomorrow afternoon. It was only a brief pause, and anyone watching would have thought it was only a passing greeting. So far everything had gone as they had planned. Tomorrow just as the Dark Force took over they had a chance to capture a high ranking Viet Cong. The Special Forces Advisor spent the day getting his PRU Tiger Scouts ready and equipped for the mission. Gilbert spent his day going over the plan, studying the charts, and preparing alternate plans for just in case.

That evening as he ate in the Officers Special Assistants (OSA) officers' mess a young First Lieutenant Infantry Officer approached his table and asked if he could join him. Gilbert waved him to a seat and continued eating. Before he had taken a bite the Lieutenant said, "I would like to join your PRU Tiger Scouts on the raid you've got planed for tomorrow." Gilbert couldn't have been more surprised if he had dropped a dead fish on the table. The mission was secret, and here was an Army Infantry Officer who he hadn't met before asking to go on the raid. "What raid are you talking about?" Gilbert asked. "The Providence Officer in Charge (POIC) told me that you were going out tomorrow. I need one more day of Combat time in the field to qualify for my Combat Infantry Badge (CIB)." The young Lieutenant answered. "This is news to me. Did he tell you when or where the PRU Tiger Scouts are going?" Gilbert asked. The Lieutenant didn't know but only that it was supposable a secret mission. The POIC was an old CIA Agent (Jeweler) who believed that contract CIA Case Officers were only uneducated fortune hunters or mercenaries. He was one of those that thought any mission where a CIA Jeweler wasn't in charge couldn't be important.

Gilbert took the time to tell the tall slim young infantry officer that any raid the PRU Tiger Scouts may go on would not be an Army mission. The people they were after weren't soldiers but were Communists and murderers. The fighting would be one on one, and Army rules of engagement didn't apply. He then went on and told him that there was no way he could ever join one of his operations just to get a medal or a combat badge.

It was early in the morning just before the sun came up when a knock came at Gilbert's bedroom door. "Come on in," Gilbert called out. It was the Infantry Lieutenant carrying a cup of tea. He was dressed in a new PRU Tiger Stripes Uniform. "I called Special Operations in Saigon and they assigned me to Phoenix Operations and attached me to your PRU Tiger Scouts," He announced smiling.

Gilbert set on the edge of the bed drinking his tea and thinking about what could be done to overcome the fact that his forth coming snatch and grab raid was now known by the Viet Cong. The Viet Cong Spy in the Special Operations Section in Saigon that Colby had warned him about must be using the Lieutenant as a Tracker to find out the target, its location, and time the raid was to take place. Gilbert even thought a minute about canceling the mission but just as quick decided to use the Lieutenant as a decoy.

"Ok, you are now a member of the team. You will be working with the Special Forces Advisor of the PRU Tiger Scouts," Gilbert answered. Then he added, "Contact Saigon Operations and tell them that you are on a PRU mission to Ninh Hoa. Ask them for three UH-1 Army Slicks to pick up our team at 1100 hours." While the Lieutenant was making the calls to Special Operations in Saigon Gilbert was meeting with the Special Forces Sergeant in the middle of the PRU drill field. They agreed to have the ARVIN PRU Chief and thirty PRU Tiger Scouts take the three Slicks to Ninh Hoa and return to arrive at 1600 hours.

The Special Forces Sergeant agreed to take command of the Tiger Scouts who were going to make the first attack and then fade away. The Lieutenant was to be with him and Gilbert would be in charge of the actual snatch and grab. Their call sign would be Fox One, Fox Two, and the reserve force in the helicopters would be Fox Three. The returning Tiger Scouts, Fox Three, would make contact by radio from the helicopters, and if they were needed the Slicks would insert them at the point where they popped yellow smoke. It turned out to be a good plan.

Gilbert hoped that the VC would believe that the raid was going to take place ten miles from where they were actually headed. Also he would have a reserve force in the air and able to reach his actual area of operations within five minutes as a back up.

The high ranking Communist was supposed to be in the target area near the hamlet at 1600 hours, and the PRU Tiger Scouts had to be in place at least an hour before the Viet Cong arrived. Hung, Gilbert's Combat Interpreter, drove the same truck that they had used on the police raid a few days before. With the Special Forces Sergeant, the Lieutenant, Gilbert, Hung, and the PRU Tiger Scouts there were fifteen men all dressed in Tiger Scout Uniforms. Their weapons consisted of M-16s and what other weapons the individual soldiers thought they needed and could carry. All the PRU Tiger Scouts also carried a machete or a long knife. When they approached the last bend in the road before the hamlet the Special Forces Sergeant, the Lieutenant, and two PRU Tiger Scouts jumped out of the truck and disappeared into the brush. The truck continued down the road about a half mile and turned off on to a trail leading away from the water that circled around back to join the main highway leading into the city. The trail was not built for vehicular traffic but was only a short cut for people walking to the highway leading into the city.

After hiding the truck off the road they positioned them selves to receive the Viet Cong who would hopefully be running away from the first ambush. At 1610 hours Gilbert received a radio call from the lead helicopter reporting that they were two minutes out. He instructed them to hold west of the highway. No sooner than he said, "Fox One out." He received a hushed call from Fox Two reporting that the target had arrived. Then he radioed, "It's a big fish." Followed by, "Fifty North Vietnamese Regular Army escorts." Gilbert radioed, "Can you make a lot of noise. Fall back and make a run for the highway?" Suddenly they could hear what sounded like a full pitched battle. Then a radio call, "The lieutenant thinks this is a real war. He won't disengage. The fish is a woman heading your way with ten soldiers. I've one casualty." Gilbert radioed, "Disengage, Pop yellow smoke. Fox Three will be there in two minutes. Our fish is in sight, out."

Hung and the PRU Tiger Scouts took up positions where their field of fire would be down the road toward the direction the enemy was coming from. Gilbert ran across the road onto the trail leading toward the water, and then threw himself down with his feet facing the dirt road. His plan was simple. When his PRU Tiger Scouts opened fire they had to kill as many of the enemy as possible before they knew it was an ambush. His PRU Tiger Scouts were out numbered, and the enemy was regular North Vietnamese Army soldiers. There would be no second chance. If the woman ran up the trail he would knock her off her feet with his legs and grab her as she fell. If she stayed with her bodyguard he would be able to attack them from the enemy's flank. One way or the other he was determined to capture this Communist. He wasn't thinking about anything else but the moment.

The woman came rushing down the trail with her escort, and had no idea that they were heading into an ambush. The soldiers were all looking back towards the sounds coming from the firefight behind them. The big fish was an older woman of about sixty-five who was leading them, and she was dressed in white silk trousers with a yellow dress split on both sides. Beside her was a younger Chinese man wearing dark pants and a white shirt carrying a leather brief case on a strap over his shoulder. Just as they passed the trail that Gilbert was laying on, face up, with his M-16 watching the road. Hung started firing with his CAR-15, and all the PRU Tiger Scouts joined in with their M-16s.

The woman turned and darted into the trail where Gilbert was waiting. The Chinese man with her fell mortally wounded just before he reached the trail. Now the killing zone was clear for the PRU to fire without worrying about hitting the woman or Gilbert. He had his hands full and couldn't hear a thing. Time seemed to stand still. When the woman had run about five feet up the trail he tripped her legs with his feet, and she fell on top of him. He grabbed her and started rolling while feeling her all over searching for weapons. The surprise was complete and as she fell her legs flew up. His hands began reaching under her dress, and into her blouse searching for weapons trying to prevent her from committing suicide. She was a strong woman and was fast becoming the aggressor. Gilbert found himself trying to protect his private parts, and struggling to keep her teeth from chewing his ears. He was no longer worried about her committing suicide, but about his own survival.

Suddenly Hung and another PRU Tiger Scout pulled her off of him. As Gilbert looked up Hung laughed saying, "If you're through feeling the lady maybe we should leave." Gilbert just glared at him and thought to himself that he was a cheeky little bastard but instead asked, "Any casualties?" Hung answered, "They never knew we were here. The man with the leather case must have been her secretary." He was holding out the brief case toward Gilbert. "Put it over your shoulder and let's get out of here. Have the men put the North Vietnamese weapons in the truck and let's get going," Gilbert ordered.

They all turned running and jumped into the truck. Hung was in the driver's seat, the woman prisoner next to him, and then Gilbert. As Gilbert climbed in he handed his M-16 up to one of the PRU Tiger Scouts and spoke to their prisoner in English saying, "If you give us any problem I'll just shoot you in the leg." As he spoke he showed her his 25 Cal Browning Automatic pistol changing it to his right pocket. Hung had his CAR-15 slung over his shoulder with the brief case. The helicopters had dropped off the ARVIN PRU Chief and the Tiger Scouts who had joined in the firefight at the first ambush site. So far no one had shown any interest in what was going on with the woman they had been escorting. Gilbert began to think that they had a chance to pull this off but then Hung started driving the truck over the walking trail, and he wasn't so sure.

The truck rolled side to side, hung up, and lurched forward sometimes almost coming to a stop but somehow it kept moving. Gilbert now had time for the radio so he turned it back on. He had turned it off during the stake out to prevent any noise from giving away their ambush. "Red One to Red Two over," He called. "Red One this is Red Two. We have three casualties. No wounded. When Red Three joined us the Lieutenant led a charge that stopped the attacking North Vietnamese soldiers. They are now in a holding mode. It looks like they think the big fish is still swimming and are buying time. We have called for Army Cobra Gun Ships. Two have reported five kilometers (kliks) out with two hunters (OH-6)." Gilbert asked, "How is the Lieutenant doing?" Red Two answered, "He didn't make it, out."

Just then the truck gave a sudden lurch, balanced its self on two wheels, and then jumped forward still upright. Hung turned the headlights on as darkness had closed in. The truck nose dipped down and then the light beams shown almost straight up as Hung drove across the ditch onto the highway. The truck was a complete wreck but still running strong as Hung turned toward the city and the safety of the National Police Compound. Hung was now driving with the throttle pressed against the floor boards. Their prisoner hadn't said a word since the ride started. Gilbert's arm was bruised from the grip of her fingers as she held on trying to stay in the seat.

When they entered the Police Compound Hung slowed down but never stopped until he arrived at the Province

Interrogation Center (PIC) building. The PRU Tiger Scouts were laughing and shouting insults at the White Mice (police). They were intoxicated with their victory and of course for all of the loot they had gotten. Gilbert's policy was that everything but the weapons belonged to the men who had killed the enemy soldiers.

After the truck stopped Gilbert dismounted and offered his arm to the lady prisoner. Somewhere between entering the highway and stopping at the Province Recon Center (PIC) building he had changed from the commander of a fighting force to a CIA Interrogator. As she used his arm for support he could feel the strength in her body. He guessed her weight at about 145 Lbs US, tall as a man (Vietnamese) with steel blue eyes, and dark hair with flecks of white. At 65 she was a handsome woman who was sure of herself but seemed disgusted with her bodyguard and showed no remorse for their loss.

Two hours ago she was to be the guest and speaker at a Viet Cong meeting with an escort of crack Communist North Vietnamese soldiers to demonstrate that the Americans were being beaten and would soon sign the surrender terms or the Paris Peace Accord. Here she was a prisoner of an upstart American who was younger than her oldest son, and the insult was that he had done it using Vietnamese traitors (PRU Tiger Scouts). Gilbert didn't know what she was expecting but it wasn't what she saw. The PIC was clean and smelled like fresh flowers. The three Interrogators had lined up to greet them as they walked in. His teaching had made a difference, and Gilbert was proud of the way everyone was playing their part.

He turned and speaking in English introduced himself saying, "my deepest apology for the way I grabbed you when we first met. I must admit I found it hard to stop. I am Gilbert H. Moriggia, Commander of the PRU Tiger Scouts who are fighting the North Vietnamese Communist." Then he introduced his interpreter Hung and the other PIC members asking, "How may I address you, my lady?" She answered, "I am known as Nuygen Nght Tu, Communist Peoples Revolutionary Party Member of this District."

Gilbert had never heard of a PRP member of this name before. He had heard of a senior hardcore communist woman who was the prime leader of the entire sapper (terrorist) in South Vietnam. This had to be the same person only she had used one of her many aliases when she introduced herself. She had lost all fear and was acting as if she was his guest. "Your secretary's briefcase was recovered, and I have to decide who gets the contents," Gilbert said. Then he added, "My policy is similar to the Communists. I return personal items to its owner and give the other items to the individual who kills or captures the enemy." The PRP Member asked, "What percent does the state take?" Gilbert answered, "I take all government papers and the weapons. Everything else is divided as I say." She thought about this for a moment or two and said, "All the money in the case is my personal property."

Hung was listening intensely to Gilbert's questions and the answers. When they had searched the briefcase it contained a copy of a speech, a plan for an attack on local targets, lists of members of the Viet Cong sapper squad, and Vietnamese Dong (Vietnamese money) needed to fund the attacks. "I am the representative of the government so the weapons and plans belong to me. You are a representative of the Communists so the money can not be yours. All the Vietnamese Dong will be given to the soldier who killed your Chinese secretary," Gilbert said. Hung was grinning from ear to ear and then said, "Dinner and the beers are on me tonight." Gilbert speaking to the Communist PRP Member said, "My lady have a good night. We will visit again tomorrow." The three PIC Interrogators were grinning as they escorted her to a cell that had a cot and a bucket in it, but no window. She looked a little dazed from the tone of Gilbert's voice for it was hard as steel and from a man who was in complete control.

That night Gilbert carefully wrote his after action report detailing the death and actions of the young Army Lieutenant. The only thing he left out of the report was that the regular North Vietnamese soldiers were escorting a Communist PRP Member and that she had been captured. Gilbert actually thought that this would keep the North Vietnamese and the Viet Cong from finding out that he was in the area.

It was a waste of time as the Nha Trang underground was abuzz with what had happened to the North Vietnamese soldiers. The Peoples Revolutionary Party Members speech was a well organized and a published event designed to show the Vietnamese people that the Americans were losing the war. Now all they had to talk about was how three American fighting men and ten PRU Tiger Scouts had walked all over the North Vietnamese soldiers sent from Hanoi for this event. They didn't know about the PRU Tiger

Scouts reserve force or of the US Army's Cobra hunter killer teams that had saved the day, and no one told them.

The next morning before Gilbert went to the PIC Interrogation building he stopped at the OSA Compound to turn in his after action report to the Company's Province Officer In Charge (OIC). Hung waited in the jeep as Gilbert planned only being there long enough to make his report. After being sent in to see the POIC the man asked, "How did the young Lieutenant do? Is that his recommendation for the CIB badge?" (Gilbert wrote in his journal that night; the POIC seemed to think that this mission was a fun trip to the beach. He has no idea what I'm really doing here.) Gilbert answered, "He didn't make it. It's all in the after action report. It was a larger force of North Vietnamese regular soldiers than we expected. He was killed leading an attack on the enemies' main force. Maybe you could see that his family gets his Purple Heart."

The POIC jumped up and leaned on his desk glaring at Gilbert. "What do you mean by engaging a North Vietnamese regular Army force? My orders clearly state that there is to be no attack or contact with the North Vietnamese regular Army. It could disrupt the Paris Peace Talks." He was clearly upset and while Gilbert stood there the man quickly read the report. Gilbert was thinking to himself that Colby knew this man and had sent him to Nha Trang for just this reason. The POIC looked up from the report and said, "These weren't regular North Vietnamese Army soldiers. No one can kill that many Communists soldiers. You will just have to rewrite or change the report." Gilbert just handed him three insignias taken from the soldiers uniforms and said, "Just attach these to the report." The POIC was now clearly upset and then he said, "This report is classified secret. There will be no Purple Heart. Give me the rest of the insignias, and I'll write a new report." Gilbert just grinned and said, "I'm sorry but they make good souvenirs, and the PRU Tiger Scouts have been giving them away."

As Gilbert turned and started out the door the POIC said, "I almost forgot but a letter came addressed to you from Mr. W. Gage McAffe in the State Department Carrier Pouch." Gilbert accepted it and remarked that it had been opened. "I open all letters in that come in from the State Department. This is my Province and no one operates in it with out my approval," the POIC answered. The letter informed Gilbert that Mr. William Colby would like to see him when he next visited Saigon. It was really a message from Bill telling him that the Viet Cong knew he was operating in Nha Trang. It was also a notice that it was time to move to his next area of operations. Bill Colby had told Gilbert that he would be organizing his type of operations in Qui Nhon and Plei Ku but nothing after these other two cities. Gilbert was sure that Colby was operating under a precise time schedule and suspicious that it had nothing to do with the State Department's plans.

Upon leaving the OSA Compound Hung drove them over to the National Police Compound where Gilbert asked the Chief of Police if he had any Viet Cong in the holding tank that hadn't been charged yet. Upon giving him an answer that he had two; Gilbert asked if he could bring the Communist PRP Member over to look at the prisoners.

Afterwards he would like three of the prisoners who were not VC released, and the Viet Cong that she had looked at to be roughed up before being placed with the other VC prisoners in the prison.

"How will you to know what man is a Viet Cong?" the Chief asked. "You will point him out but the prisoner shouldn't know that I've been told," Gilbert answered. Then he said, "I will visit you in an hour and then again in the afternoon. I'd like your police to make a sweep of an area where VC gather and capture another one but be sure to bring in at least two or more innocent bystanders with him. I'll need them for tomorrow morning. No VC women for this operation. It's got to be only Viet Cong men prisoners for this to work." The Police Chief sat back thinking about all this. Gilbert said, "Oh, I've forgot but I'll pay for all the VC captured. I'll even pay for the prisoners who you release at the same rate." The Police Chief asked, "Does that include the bystanders we pick up?" "It sure does," Gilbert answered. The Police Chief was smiling and nodding at the same time.

As Gilbert started out of the office the Police Chief said, "The tall Vietnamese with the torn shirt and with one black eye is a VC. The other VC is the one with no shoes and two black eyes." This was not the way Gilbert would have marked his prisoners for identification but it worked for the Chief of Police.

Five minutes later Gilbert and Hung walked into the Province Interrogation Center (PIC), and the Interrogators were smiling as they briefed him on their prisoner. It seemed as if she had tried to order them about and had told them that the province where she lived the interrogators were more efficient. To impress them she had inadvertently told them the name of the province, village name where she came from, and even her duties as a Communist PRP Member. "Did she name any other Communist Officers or Viet Cong leaders or members?" Gilbert asked. "Not Yet." was the answer. "What is she doing now?" Gilbert asked. "We had her clean her own bucket before giving her a can of your Army "C" rations for breakfast. She's been sitting on her cot waiting for you for the past hour," one of the PIC Interrogators answered.

Now it was Gilbert's turn to try and get the names of the key Communists and terrorist in the province. All he had to do was get the names, and the Chief of Police would do the rest. The really big problem was time. Colby had warned him that he had to move fast but it still takes time. As he walked in with out knocking the prisoner was sitting on the edge of the bed and actually seemed glad to see him. "Good morning," Gilbert said in English. He was dressed in a white shirt with a shoulder holster for his Belgium 9mm automatic pistol and dark paints. He looked like just what he was. A tough man who was in charge but also a man who wanted to please her. When she stood up she was tall enough to look directly into his eyes. She didn't talk down to this man but treated him as an equal.

He invited her to walk with him to the police headquarters saying that they could have a cold drink before she had to return to her room. She seemed eager to get outside, and as they left Hung fell in walking just behind them. Gilbert explained that besides being an interpreter he was also their bodyguard. He laughed and said, "We don't want any Viet Cong trying to attack us as we talk." They were strolling across the grounds toward the police headquarters at the time, and when they entered the building they were facing the holding tank or cage. The holding cell had ten prisoners waiting to be interviewed by the police. Gilbert pointed out the prisoners in it, and the White Mice working at work stations around the cage as they circled heading to the police break room. The break room had no wall or door but it had a bench attached to a long table for the policemen's use.

There was no one in the police squad room as Gilbert and the Communist PRP member sat down across from each other. Hung took up a position so no one would come in to bother them. Gilbert got two orange crush sodas from a cooler and as they drank said, "I will need you to give me the names of Viet Cong Officers or Leaders in this province. I need this information so I can protect you from the Viet Cong." She did not laugh but answered, "The Viet Cong people will sing songs of praise about me. They love me and nothing you can do to me will change that." "Don't look now but I think that the tall man with the black eye and torn shirt knows you? I don't think he likes you," Gilbert answered.

As they left the building Gilbert was smiling to himself. She couldn't help it but as they circled the holding tank she stared at the man. As they walked back toward the Province Interrogation Center (PIC) building two police officers came out with three prisoners and escorted them to the gate where they were released. They were told that the Communist PRP Member had identified the VC, and they were free to go. The Chief of Police then sent two policemen into the holding tank, and they beat on the prisoner giving him another black eye. The man was told that the Communist PRP member had identified him as a member of the Viet Cong, and then the police took him over to the prison to join the other VC that were being held prisoners.

That afternoon after the same short walk and visit to the police headquarters for a drink of orange crush they visited like old friends. Gilbert gave her some hard candy to eat after she was in her room but didn't ask for any names but only suggested that he would do his best to protect her from the Viet Cong sappers (terrorist). This time the Chief of Police released four men telling them that she had helped verify that they were not VC and had pointed out the man with no shoes.

After they had returned to the Province Interrogation Center (PIC) Gilbert phoned Bill Colby and told him about the raid but not what he had planned for the Communist PRP Prisoner. He also told him about the reaction of the CIA POIC. Bill didn't act surprised or really concerned. Colby asked, "Do you remember the other locations we talked about?" Then after Gilbert said he did Colby answered, "It doesn't matter which location you go to next but try and protect yourself. Those college kids at the Embassy are screaming to high heaven that you are making the Communists mad.

The Ambassador actually believes that you are a wild cannon and are making the North Vietnamese very angry.

The Commanding General for the USAF and the US Army have both asked General Abrams to give you some type of award but the Ambassador won't hear of it." Gilbert answered, "The Ambassador is right about one thing. The North Vietnamese are really pissed. Do you want me to kiss and make up?" "Hell no. If you do it will really upset me." Then he added, "I think what really made them mad was that you disappeared for three weeks, and they only discovered your whereabouts when the North Vietnamese complained about the attacks on their regular forces at the Paris Peace Talks.

Every time he talked to Bill he got the feeling he wasn't on the same page or sometimes not even on the same planet. Tomorrow Gilbert would have to put the pressure on the old lady. If he had everything figured right she may give him the information he needed to stop the sapper (terrorist) attacks in the city of Nha Trang. He knew one thing for sure. The Province Officer In Charge (POIC) wasn't happy being kept out of the loop, and it was time to leave for his next assignment.

That evening after he returned to OSA Compound and Hung had left for the evening one of the guards at the gate notified him that he had a visitor. Gilbert laughed and asked if it was a man or woman. They said it was a beautiful Vietnamese woman who said she had came from Saigon to visit him. The guards seemed impressed with her beauty and asked Gilbert where he had met her. He thought that Hung was just having fun with him because he hadn't visited any of the joy houses.

Everyone was laughing as Gilbert went out to the gate to see this beautiful Vietnamese woman who wanted to visit him. He was really surprised as it was the VC Vietnamese Doll who he had helped in Tuy Hoa, and then sent to live with her aunt in Saigon. Now Gilbert was a true if not always smart Interrogator who started wondering how she could find him when the CIA and the Viet Cong had lost track of him for three weeks. It must be true love or the Viet Cong really wanted to use him as a Tracker. She was standing there with a small bag at her feet looking as beautiful as when he last saw her in the French Mansion in Tuy Hoa.

"Come in and let's go to my room so you can rest before we have supper," Gilbert said. Then he asked, "Will you be staying long?" She was smiling and acting shy. "I must stay with you from now on," She answered. "Who said that you had to stay with me?" Then before she could answer added, "It doesn't matter. You look good enough to eat. I'm hungry. Let's talk about this tomorrow."

First it was the State Department, next the CIA, and now the North Vietnamese who wanted to keep him in sight. Gilbert thought that Colby must be an important man or some type of super patriot to ignore US politics. It didn't really matter much to Gilbert as he enjoyed his work and the challenges that went with it. The visit by the VC Doll was just one of the things he had to endure in his job. Being trained as a contract employee of the CIA working for the US Army as a Police Advisor did have its perks.

The next morning he felt alive and very good. It must have showed for when Gilbert went to the Province Interrogation Center (PIC) building to pick up the Communist PRP member she remarked that he must have had a good night. "I assure you that I was only thinking about how we could help each other that made me sleep so good," Gilbert answered. Then they walked across the compound grounds to the police headquarters to again drink their sodas in the police break room. This time she asked, "Why are the prisoners trying to hide their faces?" "Are you sure? I didn't see anyone hiding," Gilbert answered. Then he added, "Maybe you should try and see if it is someone you know? I'll wait just outside of the door." He had not asked her to give him any names since that first day. This afternoon he would ask her to name the Viet Cong leaders after their visit to the VC prison.

That afternoon as they went for their walk Gilbert told her they would skip the police headquarters because the prisoners didn't like her. When they entered the prison they walked through to the center courtyard. Gilbert told her that the prisoners they would see were all Viet Cong, and that they were controlled by someone who must be at least a Major in the North Vietnamese Army. He was a prisoner too but the National Police hadn't been able to discover who he was. Gilbert told her that he had been asked if she could point him out. Gilbert told her not to worry as it wouldn't matter who she picked out. The Communist PRP Member said, "I'll never help you. The Vietnamese people love me and will sing my praises forever." Gilbert led her out into the courtyard. The prison guards were prepared for Gilbert's visit and had all the VC prisoners in a formation in the courtyard waiting.

For the past three days the Chief of Police had the VC prisoners released into the prison thinking that they had been identified by the PRP Member. Also that she was the one who had been a secret police informer. All the prisoners who had friends and family outside the prison had being getting the same information from their families. Everyone believed that she was Gilbert's friend and that she was helping him. It had to be true for she had led the North Vietnamese soldiers into an ambush and all were killed, but she didn't get a scratch. They really hated her.

As they walked out where everyone could see her, Gilbert leaned over and whispered, "Stop here. Now we can leave if you want or maybe you should watch and see what they really think of you." As they stood waiting the police came rushing in, and grabbed the North Vietnamese Major. He was the prisoner who had been posing as a Viet Cong, and the policemen started beating him. The prisoners went crazy calling out names of hatred and shaking their fists at her. Hung jumped in front of the PRP member leveling his Carbine-15 at the prisoners as Gilbert rushed her out of the courtyard.

He stopped with her in one of the outer office room of the prison. The noise of the riot was loud, and the Communist PRP Member was actually shaking. Gilbert asked, "These are the men who will sing your praises? Do you want to go back and face them or do you want me to help you?" Gilbert asked. She only said, "May I have a paper and pencil. Then she sat at the desk in the office of the prison and wrote fourteen names of men and women with their positions within the Viet Cong organization. Gilbert then took her back to the PIC building and said, "I will visit with you tomorrow."

Then he and Hung drove over to the National Police Headquarters. While Hung waited Gilbert went in and gave the Chief of Police a copy of the list of names and the location of the major Viet Cong leadership in the Province. The Chief looked up after reading the list and said, "She turned and you have only had her for three days. What are you going to do next?" Gilbert answered, "I'm going to pay you double for the fourteen people whose names are on the list if they are in the hands of the Province Interrogation Center (PIC) by tomorrow.

Now all he had to do was figure out was how to disappear and leave for his next base of operations. First he had to send his Vietnamese Doll back to Saigon and somehow get to Qui Nhon with out being discovered. She had learned a lot about love making during the few short weeks she had been in Saigon visiting her Aunt and sure knew how to make a man happy. When morning came he wasn't sure if he was strong enough to get out of the sack, but he had promised to take the old lady to have an orange crush and for a walk as they had been doing for the past few days.

When he finely got out of bed his bed-partner was still sleeping. He wasn't quite awake when Hung drove him over to the PIC building. The PRP woman's room was empty, and this really woke him up. One of the Interrogators told him that the National Interrogation Center (NIC) had picked her up last night. He couldn't blame the Police of Chief for notifying the National Interrogation Center (NIC) as he was only trying to make points with his headquarters. This was a high profile prisoner, and the NIC always wanted these types of prisoners to interrogate The Chief had given Gilbert the three days that he had asked for, and he knew the end was coming; only the three days had gone by really fast.

Gilbert phoned Colby telling him that his prisoner had turned, and he expected to have all the Viet Cong leaders in Nha Trang in police custody or on the run by sundown. Colby told him that he had read the National Interrogation Center (NIC) Report, and that the Communist PRP member had been talking all night about the Communists Political Aspirations and Intentions. Then Colby laughed and said, "The Ambassador is upset. It seems that the PRP prisoner has been asking when you are going to take her for an orange crush. The NIC Interrogators seem to think that this is some sort of a code. Gentle Ben wants to know if you still mix it with beer."

It was time to leave Nha Trang, and he had learned a lot. The State Department was still playing the same old Political Game of gentlemen playing war with the Communists. Feed the news media with lies like Hitler did to the German people during WW-II, change morality into dollars, and hide the identity of the real enemy until after the Dark Force took over.

Gilbert left the PIC Building and had Hung drive him over to the OSA Compound where he stopped at the desk of the lady in charge of the Air America Reservations. All CIA Agents used Air America for traveling around Vietnam so they would expect him to travel this way. He asked for two seats to Saigon on a plane leaving by 1600 hours for Hotel

Man and one other (his personal code name for travel via Air America).

Next he stopped into see the POIC and informed him that he would be leaving for four days of R&R. The man knew there was a woman from Saigon visiting Gilbert, and really wasn't surprised because that's what he would do if he had the chance to go partying with a beautiful Vietnamese woman. The POIC had no idea that the girl was a VC sent to help keep track of Gilbert or would he have cared. Gilbert was working on a plan of how to disappear again.

He knew that all the Air American Flights stopped at a landing site near Buon Me Thuot, an Air American Maintenance Facility, before going to Saigon. It was located about halfway between Da Nang and Ben Tre. Beside Air America there were a lot of US Army, Marine, and Navy helicopters coming and going all day long. When Air America planes stopped in route to different destinations the passengers would off load and go into the waiting room while the planes were refueled. Then the passengers would board for the rest of their journey. This was where Gilbert planned on disappearing.

After making the reservations Gilbert had Hung drive him to the safe house. He told the Vietnamese Doll to pack her bag, and that they were going to Saigon together. Gilbert changed into a sanitized two piece Army fatigues. He rolled up the sleeves and used a piece of rope as the pants belt. From five feet away he could pass for an Army, Navy, or Marine enlisted man. He carried his 25 Cal. pistol in his left pocket and his Belgium 9 mm automatic with its shoulder holster in a small string tied bag that he carried on the plane.

With Hung driving they stopped at the National Police Compound where Gilbert went in alone to see the Chief of police. He had already captured all the Viet Cong who were on the list that the Communist PRP Member had given Gilbert except for two men. The Chief explained that the two Viet Cong had disappeared but they would be captured by tomorrow. Gilbert said, "I am going to Saigon for a party with the Vietnamese woman who came here to visit me. I will be gone about four days.

Gilbert told him to turn all the VC he captured over to the PIC Interrogators. After they have all the information that can be gotten Gilbert suggested that he may want to keep those that do not turn and become his informers into the Vietnamese Court system… "These men and women are the leaders of the Communist sapper (terrorist) cells and will be a real threat to the people." The Police Chief just frowned until Gilbert said, "I stopped to pay you for the VC you captured, and I'll also include payment for the other two as they will be arrested before I return." The Police Chief's face was all smiles, and he just beamed as he said, "I think that this should all be done without bothering the POIC. What do you want Mr. Hung to do while you are gone?" Gilbert was smiling as he said, "Mr. Hung is a very rich man. He may want to retire. I think that you maybe right not to bother the POIC about little details." As Gilbert left the building the Chief walked out to meet the Vietnamese woman he had heard about. He whistled and said, "Have fun in Saigon."

Hung stopped at a restaurant he knew about on the way to the airport, and while they were eating Gilbert suggested that he might want to think about taking a trip abroad. Just before getting into the Air American Plane Gilbert told his Interpreter to turn the jeep into the motor pool at the OSA Compound. The Vietnamese Doll had no idea that this was the last time she would be with Gilbert. When Gilbert made himself disappear the second time the Viet Cong must have decided that she had assisted him. A week later the Saigon police found her body in an alley off of Tu Do Street. Her head had been cut off and was found next to the hotel. The police report stated it looked as if she was running trying to escape when the machete had done its job.

After the plane had landed at the refueling stop on Air America's Maintenance airport Gilbert and its only other passenger the Vietnamese Doll went into the waiting room to wait until they were called. Gilbert had carried in his small bag saying; that he didn't want anyone to steal his weapons. When the flight was called for them to board Gilbert told her that he had to use the latrine and would meet her in the plane.

He left the building walking in a hurry heading toward the small radio control tower. This is where the military soldiers waited for a flight. The crew chiefs for the helicopters would stop to ask if anyone was going their way, and that was the only procedure needed to get a ride. As he approached the tower a crew chief asked if anyone wanted a ride to Kon Tum. It was an Army UH-1 helicopter, and Gilbert was aboard the helicopter and had started to disappear before the Air America plane had even started its engine.

CHAPTER SEVEN

When Gilbert ran to stand at the base of the Air America Radio Control Tower next to their hanger he was looking for an Army helicopter ride to any place in any direction. His plan was to merge into the Army and disappear for a few days to break his trail for anyone trying to track his movements. The Air America Base was a dirt strip 3000 feet long with a small hanger, waiting room, refueling tanks for aviation fuel, and 500 gal rubber blivets of JP-4 jet fuel for helicopters. The airstrip paralleled the highway just west of the village of Ban Me Thuot and Hwy 21 which joined with Hwy 14 in the village before running north for a mile where they again split. The airstrip was actually located on the Plateau du Darlac where the land was flat. There were two large mountains a few miles to the east, and Hwy 21 ran east through the Lang Bian Mountains to Hwy 1. It was a beautiful twenty-five minute scenic flight from the airbase to the coast of the South China Sea.

As the helicopter took off it was starting to get dark. They were flying at about 4000 feet above the ground heading north while the setting sun was showing highway 14 as a silver ribbon stretching north beneath them to the Kon Tum firebase. The land was flat with rolling fields of green grass. The grass was slowly turning black from the creeping shadows of night but the setting sun still lighted the highway making it a silver path they could follow. Highway 14 joined the Ho Chi Minh Trail in the village of Kon Tum, and as they passed over Plei Ku there was no traffic or buildings with lights to take away from the beauty of the land. Gilbert felt at peace with the world and safe for the first time in weeks. It was the time of day when the Bright Force was relaxing, and the Dark Force hadn't yet taken over the Garden of Eden.

As they flew over Plei Ku Gilbert could see that there was a 6000 ft airstrip east of the city and that the town was built on both sides of the highway. In the center of town there was a junction for Hwy 19 that traveled east to join Hwy 1 along the coast just to the north of Qui Nhon city. No wonder Colby wanted him to cut off the Viet Cong and Communists access to the area. It was the fastest route for the North Vietnamese Army to move on if they wanted to trap the American Army when it tried to leave Saigon. If the Communists could trace Gilbert they would soon discover that Colby had effectively added two weeks to the time it would take them to surround the American Army and trap them in the III Corps, Saigon area.

The Ho Chi Minh Trail started in Dong Hoi North Vietnam where the railroad and Highway 1 had truck terminals where they could off load military supplies and head inland away from the South China Sea to Laos, then south to Cambodia to avoid the guns of the US Seventh Fleet. The trail crossed the Ya Krong Bolah River as it entered South Vietnam and then intersected with Hwy 14, joining with Hwy 19 and 7 leading to the coast were they again connected with Hwy 1. All of the cities where Colby had assigned Gilbert were major hubs in the transportation route for the North Vietnamese Army moving south. The Communist had recruited Viet Cong in these transportation hubs that they needed for their master plan of destroying the US Army. They had organized them into VC Units of Sappers (terrorist) and para-military teams to hold the bridges, railroad, and crossroads until the North Vietnamese Army units could take over. Gilbert was setting up the program that was destroying the Viet Cong organizations in all of these areas. He had only started to guess at what Colby was doing, and it fit in with his desire to save the US Army soldiers from facing a Vietnamese Dunkirk.

As the helicopter was approaching the US Army Firebase at Kon Tum Gilbert showed his CID (walking on water) ID card to the aircraft commander and asked him if they had a place where he could bunk for the night. The Army aviator just nodded and went back to his landing check list. After landing, the pilot invited him to the Battalion's small Officers' Club where he met their Battalion (Firebase) Commander. The Lt/Colonel invited him to stay as long as he wanted and also offered to let him use one of his Warrant Officer Pilot's cot who was on R&R in Bangkok, Thailand. Gilbert spent the rest of the evening buying beer for his new found friends.

The next morning, after eating breakfast on paper plates in the Officers Mess, Gilbert took a good look at the firebase where he planned on staying for the next two days. It was a little more than an expanded firebase and was surrounded by a Green Line. It was a mound of earth ten feet high with concertina wire, claymore mines, and machine gun posts facing out towards plotted killing zones. The Green Line was broken by two gates in one opening that faced the village of Kon Tun. One gate was guarded by ARVIN Soldiers and the other by US Army soldiers wearing Four Leaf Clover shoulder patches (4th Infantry Division). All

Vietnamese who entered or left the firebase had to pass through the gate and be searched by the ARVIN Soldiers. All Americans passed through the gate guarded by the US Army soldiers.

The firebase also had an attached 105mm artillery battery with five tubes (Cannons) located around the base and four line companies of infantry each protecting a section of the Green Line. There was an HQ & HQ Company and a company of ARVIN Infantry soldiers in reserve. There was one helicopter pad located within the firebase with one UH-1 Slick assigned to the Battalion Commander. This was the helicopter that Gilbert had arrived on. There was also a large helicopter pad located just outside near the gate for supply helicopters that came in and left daily. In the center of the firebase was the Combat Control Center with the antenna farm and a radar artillery tracking disk.

A Battalion road ran around the inside of the compound beginning and ending at the gate. All the line companies were located between the road and the Green Line. The artillery batteries were located between the line company areas. On the inside of the circle was the Battalion Hq & Hq Company, the officers quarters, officers latrine, officers shower room, the ARVN quarters along with a small Chapel, battalion mess hall, officers club, NCO club, Vietnamese massage parlor, Korean tailor shop, Vietnamese gift store, Red Cross center, US Post Office, and a small Post Exchange (PX). In the dead center of the firebase next to the Command Fire Control Center was a one seat latrine painted white with a half moon cut into the door. It had a sign hung on the door that read, "Colonel's Crapper". All the other latrines were located near the Green Line plus each company area had a shower room.

Kon Tum was in a string of firebases located close enough to each other to provide artillery fire support in case of an attack from the North Vietnamese. The only attacks that this base had over the past year was from the Viet Cong who would fire one or two mortar rounds into the base from Hwy 14 or from within the city. The attacks normally came during the day when civilians or children were around. The Battalion Commander had retaliated by putting sandbags around all the buildings except for his outhouse. He wouldn't believe the Artillery Officers who told him that the reason the VC never fired at his outhouse was because it was being used as an aiming point for the gunners firing the mortars at other targets.

That first evening Gilbert had bought enough beers for the Lt/Colonel that they had became friends, and when the Battalion Commander asked what kind of work Gilbert did for the Army he answered, "I am a National Police Advisor. I help them solve mysteries." The Lt/Colonel then asked him if he could help solve a mystery that had been bothering him for the past five months. Gilbert agreed and asked him to tell about the problem or mystery. All the officers stopped talking and gathered around to listen. Each wanted to explain what he thought was happening. The Lt/Colonel said, "Every week day three or four young women would be seen running through a line company's area at night. This happened every night during the week but never twice in the same company's area. When ever the guards chased them the girls would just disappear. The next night they would show up across the compound in another company's area." Everyone started telling about what they had seen. One said that it was different girls each night. Another disagreed but they all agreed that they were young Vietnamese women. Gilbert was sure he knew who they were and what they were doing but he still needed to see them get into the firebase before telling the Battalion Commander how it was done. The officers all agreed to meet in the officers' club the next night, and Gilbert promised them an answer.

Through Gilbert's technique of asking the right question mixed with an unlikely one that he was able to find out that it was really young women around the age of fourteen who were coming into the firebase at night. They had small breasts and waists. They were also very fast runners and none over four feet tall. Gilbert had already guessed that they were farm girls whose fathers and mothers had sent them out to make a little extra money for the family.

The next day Gilbert sat near the front gate just watching the guards. He was looking for a vehicle that someone weighing less than a hundred pounds could hide in. It would have to be large enough to carry at least three girls and not be obvious to the guards at the gate. He noticed that a Vietnamese water truck was driving in through the Vietnamese side of the gate. It was stopped and searched before it was allowed to drive onto the base, and then it proceeded to one of the line companies shower room to fill its water tank. He watched as the guards looked underneath the truck, in the cab, one even got up on top, and peered into the opening where the water was poured into the tank when it was being filled. Gilbert wandered off after the truck had made three round trips.

He then went to visit the Korean Taylor shop to get the civilian clothing he had ordered to replace the clothing he had left at Nha Trang. Gilbert wanted to be dressed as a CIA Interrogator when he arrived at Qui Nhon so none would guess how he had gotten there. The shop was across the street from a line company that had not yet had a visit from any girls that week. He visited with the tailor while waiting until the water truck came and drove back toward the company's showers. Gilbert then left the shop and followed the truck walking between the single floor barracks but staying close enough that he could watch but not readily be seen. The driver connected the truck's water hose to the shower's water tank, and then the driver climbed up on top to open the round cover closed the water tank after it was filled.

Then the driver climbed down and started the water pump that would fill the shower's holding tank. The water pump's engine was loud enough that anyone wanting to speak to the driver would have to walk to the other side of the showers. As Gilbert watched one girl's head popped up out of the top of the truck's water tank, and then she squeezed thru the small opening. All she was wearing were the white silk slacks that Vietnamese woman wore under their dresses. She turned and helped another girl who popped out, another climbed out, then another until four young women were laughing and climbing down from the truck. The driver never looked up but continued watching the gauge on the water tank. One of the girls reached under the tank and pulled out a sack. She reached into the sack and took out four dresses of brilliant colors with split sides. Then she handed out four pair of shoes. They were high heel shoes made out of wood and painted in black lacquer with bright yellow flowers.

As Gilbert watched they quickly dressed and darted into the shower room. A few minutes later they once again appeared, and put the empty sack back in its place under the truck's water tank. They had put their hair up, and were wearing the high heal shoes making them four inches higher. If he hadn't seen them come out of the water tank he would have believe that they were at least nineteen years old. The silk slacks had dried and the girls actually sparkled in the hot sun. Gilbert knew right then that he couldn't tell the officers at the club what he had discovered. He would have to meet with the Battalion Commander for a very private talk as soon as possible. He was going to suggest that the Lt/Colonel issue an order increasing the size of the truck's water tank opening to make it possible for older girls to fit into the tank.

Gilbert knew from his experience in other cities that sex was a real problem in Vietnamese social behavior but not necessarily a security issue. If older women and girls had the opportunity to take over the trade the younger girls would not be pushed into prostitution. It was a problem but not one that Gilbert was going to discuss with drunks at the officers' club bar.

After picking up the clothing and things he had purchased in the Taylor Shop, Gift Shop, and Post Exchange he went over to the Battalion Commander's office. Gilbert was invited to visit, and he asked, "May I close the door?" The Lt/Colonel nodded and asked, "What is the problem?" Then Gilbert told him what he had discovered. The Battalion Commander told him he had guessed that it was something like that. Then Gilbert told him the ages of the girls, and the Lt/Colonel looked disgusted.

Gilbert then told him why the young women couldn't be over fourteen years old, and suggested that he order all water trucks specifications to be changed to have a larger opening made for filling the water tanks. Gilbert went on to other things that might be done. He never told the Battalion Commander what to do but only suggested security procedures that might help. Gilbert was practicing interrogation in reverse. He wasn't seeking information but trying to give it without the subject knowing what he was doing. It was a strange procedure for Gilbert.

After they had talked for at least an hour Gilbert said, "Tonight at the Club I'm going to tell everyone that I'm stumped. It's a complete mystery and I will buy the drinks. I have to leave tomorrow." The Lt/Colonel said, "You know I have a daughter that age at home." Then he said, "Thank you. Can I help in any way?" Gilbert answered, "I could use an Army barracks bag for the things I've purchased in the Post Exchange (PX). I will also need to find a ride to Qui Nhon." The Lt/Colonel said, "No problem, I'll have my pilot drop you off tomorrow morning when he makes the flight to pickup some of our men returning from R&R." Then he opened the door of his office and told his Sergeant, "Get Gilbert a barracks bag from supply and put it on his cot." Then he said, "I'll see you tonight at the club."

That evening after everyone had a good laugh at Gilbert's failure to unravel the mystery of the running girls one of the Army Warrant Officer aviators asked, "Have you ever met a pilot known as Gentle Ben in your travels?" Gilbert was flabbergasted for just a moment and then answered, "I met him once in Naha, Okinawa in 1949 when he came into the parachute shop to replace some survival gear. It seems that someone had stolen some things from his locker when he was missing for two days in his P-61 (Black Widow Night Fighter) over the Pacific Ocean. I think he was then known as Paper Doll." Another officer spoke up and said, "I was stationed in Saigon as an adjutant once when my office had to process thirty-two reprimands and one Army Commendation for Gentle Ben from Commanders throughout the III Corps Area. He had put thirty-two senior master sergeants into a hot firebase. They were aboard his CH-47 Chinook helicopter when a firebase he was flying over came under attack. The firebase commander requested all helicopters with troops aboard to help strengthen his green line. Gentle Ben was the first helicopter to respond."

Gilbert couldn't help himself and asked, "What happened?" The Officer laughed and said, "I believe that we issued seven Silver Stars, and 25 Bronze Stars plus everyone got the Infantry Combat Badge. The fight lasted for two days before the firebase commander would request UH-1 Slicks to come in and take the Sergeants back to Saigon. Those Master Sergeants never did forgive Gentle Ben." Everyone started telling stories about Gentle Ben. One even claimed to have met him flying an H-19 helicopter for General Ward during the Detroit Riots. Gilbert finely asked, "How do you know about Gentle Ben?" "We get copies of the

Army Times. There is a story about him on the front page of the Times dated 6 November 1969." "We just got ours last week," someone answered as the paper was handed to Gilbert. One time Gentle Ben had told Gilbert that there was more than one way to disappear, and the only way to keep a secret was not to have one. This was his way of dropping out of sight.

The Battalion S-2 Officer asked, "Has everyone heard that Gentle Ben was sent to Saigon Special Operations to be court marshaled for his latest escapade?" No one had, so the S-2 Officer told the story. It seems as if a 1st Cavalry Division Battalion Commander had his Battalion on the airstrip at the base of Song Be Mountain preparing to attack the mountain from all four sides and at the same time have a company of troopers come down from the top. Gentle Ben was returning from a mission of inserting a Vietnamese Scout Team across the river into Cambodia and knew that the river had flooded up to the base of the mountain. He was charged with stealing one of the Battalion's Line Companies along with its officers. It seems that he got four other Chinooks from "B" Company, 228 AV Battalion, 1st Cavalry Division to assist him.

They dropped the Company of troopers into the water at the base of the mountain across from Cambodia, and then used the Chinooks as Swift Boats to give the troopers fire support while they built platforms or fox holes above the flooded plain. The Chinooks floated around amongst the working troopers, and used their M-60 machine guns to recon by fire all the patches or bunches of bamboo before the troopers moved in. When the North Vietnamese soldiers tried to escape from the mountain they had to wade in water up to their arm pit carrying their weapon over their heads. The troopers had a duck shoot. There were no prisoners taken, and over two North Vietnamese companies bit the dust or mud. The troopers had no casualties, and when they ran out of ammunition used their M-16 as clubs.

Gilbert couldn't help himself and asked, "What happened to Gentle Ben?" It seems that the 228th Aviation Battalion Commander Lt/Colonel Blanchard at the pretrial hearing told the board that Gentle Ben was acting under his orders. The Commanding General relieved Lt/Colonel Blanchard the following day and assigned him as the new Division Combat Operations Officer. He was given a battlefield promotion to Colonel. Gentle Ben was promoted from CW-2 to CW-3 before the hearing was concluded, awarded the Bronze Star, given 30 days R&R with his wife in Saigon, and transferred to "E" Battery of the 82nd Artillery, 1st Cavalry Division. The notes I read indicated that the General's remarks to the board were very clear; it was this or a court marshal and no one thought he was talking about Gentle Ben," the S-2 Officer said.

Now Gilbert felt really alone. All the people who had started out with him were now out of his reach. It was like Colby had deliberately cut him off from all ties to the Company. A few months ago at his last CIA meeting in Saigon, and during a party that evening Caesar Civitella had told him that no matter what he did there was no way that he would ever become a Jeweler (Full Time Company Agent).

Gilbert had great respect for Caesar and knew him to be an honest, loyal, CIA Officer who had started out with the

OSS during WW-II. He was now a superior Vietnamese Province Officer In Charge (POIC), and was only two years older than Gentle Ben. The only difference that Gilbert could see between them was that Caesar worked within a structured organization, and Gentle Ben was a hands on man working on assignments without structure. Caesar had been drinking a few beers so Gilbert just ignored the man's remarks, but then the more he thought about them he wasn't so sure it was the beer talking. He decided that the best thing for him to do was to complete the job of helping Colby slow down the advance of the North Vietnamese Army and forget about becoming a CIA Jeweler.

Caesar's remarks still depressed him so he left the Duc Hotel and headed downtown to the Sportsman's Bar. It was where the Special Forces officers and men hung out when they were on R&R or during visits to Saigon. Colonel Joe Callahan, Major Jerry Walters, and their 1st Sergeant were there visiting with old friends when he arrived. It was like a class reunion and Gilbert could relax and forget about his work for a few hours. He had worked with many of the Special Forces Advisors in the field, and once in the same Province where Colonel Callahan was the Senior Army Province Officer. He soon forgot about Caesar's remarks.

Curfew had come and gone when two White Mice (Vietnamese Police) and two burly MPs (Military Police) came into the bar. One of the MP's said, "Let's see some ID's. It's after curfew." Gilbert spoke up and said, "It's Ok, they are with me. They are helping me search the bar girls." The White Mice just stood near the door watching as Gilbert handed his CIA ID card to one of the MPs. The MP looked at it and said, "Thank you Sir! Have a nice evening. I'll have the Vietnamese Police stand guard outside so you will not be disturbed again."

Colonel Callahan asked, "What kind of ID Card is that?" He snatched it out of Gilbert's hand and read it very carefully. Then he asked. "How do I get one of these cards?" Gilbert explained that all he had to do was join the Company. Every one laughed and the party went on until dawn.

While Gilbert was thinking about visiting his friends and the fun things that can happen in a city at war he fell asleep. A US Army firebase sitting astride a major supply route for the North Vietnamese Army is not exactly the safest place there is, but Gilbert felt comfortable and slept like a log.

The next morning he walked to the helicopter's log pad and met the crew as they were getting ready for take off. Gilbert noticed that there was a chain link fence on telephone poles placed fifty feet back from the pad on the side toward the green line. The chain link fence's base actually started ten feet above the ground reaching up to the tops of the poles for the length of the field. It was a strange sight, and he asked the crew chief what it was for. The Crew Chief answered, "The Viet Cong keep trying to hit the helicopter with rockets. The fence doesn't stop the rockets but it causes them to explode when they hit it. The shrapnel from the rocket falls to the ground before they hit the helicopter and that's good." Gilbert agreed.

It was a beautiful flight down from the plateau through the mountains to Qui Nhon and the South China Sea. He was still dressed in Army Fatigues, carrying a barracks bag, and when he left the helicopter no one even glanced at him. Gilbert went into the operations building and used the latrine to change into a white shirt; dark pants, and then placed the shoulder holster over his left arm. He now looked and felt like he was ready to meet the CIA Province Officer In Charge (POIC) of Qui Nhon.

This Province was organized like most of the other places where he had worked. The MACV Compound had the Vietnamese minor government offices, the US Army Senior Army Officer's office, and the ARVAN Chief of the Provenance. Everyone in these offices thought that they were in charge of the Province. Then there was the Officers Special Advisors (OSA) Compound that had the office of the CIA Province Officer In Charge (POIC) who knew he was in charge of the Province and the PRU Tiger Scouts. Separate but near by was the National Police Compound with the Province Interrogation Center (POIC).

This POIC came out of the same mold that all CIA Jewelers came from. He was good but very protective of his turf. Gilbert had hitched a ride from the airport to the OSA Compound on the first Army truck headed in that direction. He walked into the building where the POIC Office was located, and after waiting thirty minutes was invited in. Gilbert introduced himself and the POIC said, "It's about time you got here. The Province Recon Unit (PRU) Advisor left three days ago. I haven't time to do your work." Gilbert knew that the man hadn't even listened when he introduced himself. The POIC thought that he was a replacement for the man who had just left. This was a chance for him to operate with out anyone in Saigon except Colby knowing he was here. Colby wasn't expecting him to report in so he just saluted and left the man's office. The POIC didn't even look up but Gilbert had saluted hoping the man would think he was Special Forces, and maybe he did.

When Gilbert went over to the PRU Tiger Scouts' Barracks he met the ARVIN PRU Chief and asked him in English to form the men into a formation for him to check their weapons. He had decided not to speak Vietnamese and to play the part of a green Advisor. He didn't want to get too friendly with the PRU Tiger Scouts as he knew from his last assignment that their names had been sent to the Paris Peace Talks and turned over to the North Vietnamese. They were to be executed as soon as the Americans left Vietnam. There would no longer be any more VC recruited to join their ranks under Gilbert's watch.

The PRU Chief spoke a little English and with the help of two other Tiger Scouts they figured out what he wanted. After they formed up Gilbert went down the line and pushed the men around until he had them in six man squads. He had sorted them so that no man was of the same height. There were one or two as tall as he was with another shorter and one or two taller. He wanted to fade in and not stand out when he was amongst them.

He then had them fall out of the unit formation, and when he whistled they would form into the same squads again. It took awhile but they soon caught on and thought it was some sort of game. He then held a weapons inspection, and once again they found out that he was more than a man playing games. The weapons were the M-16 but not all were automatic, and he had other issues about the weapons.

Gilbert asked if there was a Company warehouse on the airbase. This was almost an impossible question for them to understand, and it took six men plus the PRU Chief working together before they understood that he wanted to know where it was, and who had the key.

The key was kept by the Province Officer In Charge (POIC) but one of the PRU wives who worked in his office knew where the key was kept. When they realized that Gilbert's plan was to steal the key and make a copy they went to great lengths convincing him that they would give him a key tomorrow. Gilbert had to keep from laughing but somehow they had become his conspirators, and he had them all working for him. Gilbert still had to find a room to sleep in so he asked the PRU Chief where the old Advisor room was.

They tried to convince him that the old Advisor slept with one of the PRU wives, and when that didn't work, they offered him a bed with the single men. Once again it was hard to keep from laughing as he listened to them plan on how to fool him. Finally he agreed that he would sleep in the barracks but all the men would have to move and sleep outside. When they realized that he was going to make them sleep outside in the rain one squad motioned for him to follow them. They led him to a room in the Officers Advisors Quarters, and showed him a nice room that was vacant. It was just what he needed. It was already on the POIC books as the PRU Advisor's quarters, and until Saigon MACV actually sent someone to replace the man that had left he had a home.

The next morning Gilbert ate breakfast with the other American Advisors to get better acquainted, and to help identify himself as the new PRU Advisor. Gilbert walked out of the mess hall with two other Advisors, and seeing some of the Tiger Scouts walking toward their barracks area; Gilbert whistled. All the American Advisors stopped to watch as the compound came to life with men forming into six man squads all over the compound. Each squad formed and went into a defensive position and waited. Gilbert whistled again, and the squads moved to positions that divided the compound into three killing zones. The Senior Army Advisor was an Infantry Officer and he asked, "How long have you been here?" Gilbert answered, "I came on an Air America flight yesterday."

Gilbert whistled again and walked over to the PRU Tiger Scouts as they again formed into a unit formation. The PRU Chief walked up smiling, and held out his hand that was holding a key. Gilbert accepted the key and ordered everyone to board the PRU Trucks. There were fifty-six Tiger Scouts that got into the pickup and a 2½ ton truck. Gilbert got in the drivers seat of the pickup, and everyone got aboard before he ordered them to follow him to the airbase. At the gate he showed his CIA ID Card, and the Air Police (AP) waved them through. The PRU Chief directed them to the CIA warehouse, and Gilbert got out of the truck. He placed the key in the lock and it fit.

He had them form up, and then he whistled. After the Squads were formed he had each group go into the warehouse and had them leave their M-16s. He gave each man a new M-16 with a CIA clip of thirty rounds. Every man was also given a machete or a long knife. Two of the squads were each given an M-79, and all the ammunition they could carry. He had every man pick out a new Tiger Scout uniform and one for himself. Then he had them pick out a new black pajama uniform and new boots with one uniform for himself. He gave each man a new canteen and a back pack. He had the men load the truck with cases of "C" rations and cases of ammunition. Then they headed back to their compound.

As they passed the US Army UH-1 Slick helicopter company stationed on the airbase he had the trucks stop and went in to talk to the Operations Officer. He showed him his CID ID Card and told him that he wanted to meet the men who were going to insert them on special hard missions from time to time. Gilbert made a point that he would like to buy the aviators a beer to get better acquainted. What Gilbert really wanted was for the Army aviators to accept his requests with out going through Saigon Special Operations. Every request that went though the Tactical Operations Center (TOC) in Saigon somehow got into the Viet Cong hands. The Operations Officer was impressed when he saw that Gilbert had a platoon of PRU Tiger Scouts as his escort. They were a hard and mean looking bunch that looked as if they would like nothing better than to kick some ass.

After they returned to their Compound Gilbert had the PRU Chief take him to the National Police Compound in the Pickup truck. This was where they found out that he expected one Squad to accompany him where ever he went. They were also surprised when Gilbert had the PRU Chief ride in the cab with the driver while Gilbert rode with the men in the back. This was only Gilbert's second day in

Qui Nhon, and now he was ready to meet his Province Interrogation Center (PIC) Interrogators.

Even before he walked into the building he was mad. There was a wooden cage out side of the building with a female prisoner in it. She was filthy and looked like an animal. There was no bucket or water near the cage. He knew it was a woman because she had no clothing on. Gilbert could smell her when he got out of the truck. There were five PIC Interrogators waiting to greet him. Gilbert was very polite when he asked, "Can anyone here speak English?" One interrogator answered that he could. Gilbert didn't raise his voice but he never asked any questions either only gave orders. "Give that girl fresh water and feed her. After she has eaten see that she is cleaned up. Let her use the shower, and get her some clothing. One of the smaller Tiger Scouts snickered. Gilbert turned and told him to give him his Tiger Scout Uniform. The man took it off and Gilbert turned while handing the uniform to the Interrogator who spoke a little English said, "Have her wear these."

The PRU Tiger Scout just stood in his altogether with a cloth that he used instead of under shorts to keep his privates from bouncing when he ran. His face had a blank look as if he didn't understand what had happened. Gilbert walked over to the truck and reached into where he had put his new uniforms. Gilbert handed the man his new Tiger Striped uniform and told him that he could wear it. The man stood looking straight ahead and said, "No, I have another uniform in the barracks, and you need yours so you can be one of us." Now Gilbert understood him, but couldn't answer without letting them know he spoke Vietnamese, so he just nodded.

They waited in the shade under a tree until one of the Interrogators came and asked him to come with him. He went with the man, and the prisoner came out dressed in the Tiger Scout uniform to stand in front of Gilbert. The prisoner was a young nineteen year old Montagnard woman who according to the Interrogator was a known Viet Cong. She had been captured the day that the old American PRU Advisor had left for Saigon. No one knew what to do with her as she was Montagnard and hadn't been charged.

Gilbert was going into his Interrogator mode when he told the Province Interrogation Center (PIC) Interrogators to have her wait while he inspected the PIC building. After Gilbert returned from the inspection he told them that the girl was to be put in the room that had a door that could be locked. He instructed them to put a cot in the room and let her use the bathroom. Gilbert also informed them that tomorrow he would bring her some shoes.

All this time the prisoner had stood and said nothing but Gilbert had noticed that her eyes were watching his every move. She stood looking straight ahead at the wall, and as the interrogators were moving about Gilbert had moved so he was beside her looking out of the room and down the hall. Her soft voice in perfect English asked, "Do you not want them to know you understand Vietnamese?" "Yes" He answered without looking at her.

That evening Gilbert went to the small Officers' Club for a beer and to let his fellow Americans accept him as a new member of the Advisors. He wasn't trying to fool them but only to protect himself from the results of the raids he was going to be carrying out against the Dark Forces starting tomorrow. He just needed a target, and that would come from the first VC they captured in their snatch and grab raids.

From the information he was getting Gilbert thought that there was at least seven weeks left before Colby would be forced to recall him, and the opportunity to delay the North Vietnamese Army would be gone. He couldn't follow Colby's time schedule if he was going to be successful. Gilbert came up with a plan known only to himself. He would use the triangle of Qui Nhon to Plei KU to Kon Tum and return as his area of operations. This way he would have a base at each location and could move fast with out the enemy knowing where or when he was going to hit them. If he moved fast enough not even Colby would be able to guess where he was going to be next.

Colby had told him once to ignore all orders except his, and that Gilbert was to agree to everything he said or instructed but not to stop his attacks. This was the Army's only chance of avoiding a retreat like the British did at Dunkirk, France, and even they couldn't guess how close it was going to be.

The next day Gilbert instructed the PRU Chief to have all the Tiger Scouts wear the Striped Uniforms during the day and after sundown to change into the Black Pajama Uniforms. He allowed them to wear the Black Hat with both uniforms. He also instructed the PRU Chief to have the men practice firing their weapons, and that he would inspect them in the afternoon to see how good they were. Everyman who could meet the requirement of sharpshooter would be given a cash bonus of twenty-five Dong that very day. Gilbert knew he had to have everyone at their peak and motivated. With these men it was money first, last, and always.

He picked out a Squad of Tiger Scouts, and told them to get in the truck as they were going to the Province Interrogation Center (PIC). It was the only time Gilbert almost had his orders disobeyed. They wanted to stay and practice with their weapons. It was only after he promised that they could practice after returning was he able to get them on the truck. Gilbert drove directly to the National Police Compound, and upon picking up the prisoner they headed out to the airbase for a visit to the base dispensary. He spoke to the Montagnard girl in English instructing her to tell the men to guard the truck and wait. Then he walked inside. She followed dressed in her Tiger Scout Uniform wearing the dark sunglasses and the new boots he had brought with him to the PIC.

Gilbert asked to see a doctor, and showed him his CID ID Card while telling him that Kim was his Interpreter and a PRU Tiger Scout. He also told him that she had been a prisoner but not who's prisoner, and that he would like a complete physical performed to be sure she could continue with her duties. The doctor agreed, and then Gilbert told him that because of the classification of her work no Vietnamese were allowed to speak or be near her at any time. "Will you be present in the room?" the Doctor asked. "That is not necessary. Just be sure you have a female nurse with her at all times," Gilbert answered.

When Gilbert had picked her up at the PIC he had asked the Interrogator if she had eaten. When told that she had; he then turned and explained to her that they were going to an American Hospital to see if she was fit to help him. She had nodded but when he gave her the black wide brimmed hat, boots, and sunglasses she had started to smile and had asked, "What are you expecting me to do?" Nothing really but I don't want anyone to recognize you," Gilbert answered. Then he added, "I will call you, Kim. That is both a boys and woman's name. If you will act as my interpreter I will pay you as a PRU Tiger Scout."

She had not spoken all the way to the airbase, and he took a chance letting her go with the doctor alone. Gilbert was building a small but deadly fighting force, and even he wasn't sure if it would work. The PRU Tiger Scouts were trained and organized to capture but not kill the Viet Cong. The old rules were changing, and now they were going to have to fight. The PRU Tiger Scouts and their families were all on the death list of the North Vietnamese Army only they didn't know it. It is true they were Buddhist but that meant nothing to the Communists. The main reason he was using a disguise for Kim was that she wasn't on the PRU list that was given to the North Vietnamese at the Paris Peace Talks, and with luck she could disappear when this was over.

As she went into the examining room with the doctor Gilbert asked one of the nurses if he could get her to obtain some feminine items that a women needs. He told her he would pay whatever was needed, and that Kim had been a prisoner who had escaped with nothing. He told the nurse that under clothing would be nice plus any toilet articles available. Gilbert handed the nurse a hundred dollars in American Script when she told him that the AFX (Air Force Exchange) had the items.

Gilbert left the building and joined his PRU Squad, and listened as they told about their conquests with women. When PRU Tiger Scouts talk after a mission or on a break it sure wasn't bragging about killing someone. It was about women or money. They were talking about the twenty-five Dong bonus they were going to get and what they were going to do with it. He had never realized how much he could learn by just pretending that he didn't speak Vietnamese. They sometimes referred to him but it was with respect or of how they had tried to fool him, and he had caught them. They were like deadly children.

It was getting close to noon when Kim came out of the examining room with the doctor. He told Gilbert that she was in good health but could use a few more good meals. Gilbert laughed and turned to leave while Kim carried a sack and three bundles wrapped in paper. She was smiling when she said, "The nurse gave me these gifts that you ordered. I will be your Interpreter." Then she asked, "What else will I have to do?" "You will have to cut your hair like the Tiger Scouts and learn to shoot like them," Gilbert answered. All she said was, "OK".

Gilbert started thinking how he was going to feed them. He decided that he would go to the Air Police Mess and get box lunches. They could eat them at the CIA warehouse while he picked up some more ammunition and "C" rations. Gilbert had also decided on giving Kim another uniform so she could have it tailored to fit better. She had to look like a Tiger Scout when they were in the field and to handle weapons like a pro so as not to bring attention to him or herself.

After they left the CIA Warehouse Gilbert stopped at the US Army Helicopter Company and asked the Operations Officer if he had a helicopter heading towards Kon Tum late today. He would like a ride for his Squad to the firebase. The Company Commander was visiting his Operations Officer and when Gilbert asked for the ride. He introduced himself asking, "Is this a mission assignment?" Gilbert answered saying that he was friend of the Battalion Commander of the Kon Tum firebase and had been invited to a party at the officers club. The Tiger Scouts were his bodyguard, and his friend would give them a ride back. The Major told him that the Battalion Commander was an old friend and had graduated in the West Point Class ahead of him. Then he turned and told the operations officer to find someone to take them to Kon Tum.

On the way back to the PRU Compound they stopped and Kim picked up all of her things from the PIC. It was only the few toilet articles that Gilbert had given her the day before but Gilbert wanted the Interrogators to know that she was now working for him. He also gave them the money to paint the building inside and out. Then he told them that he was going on a snatch and grab raid to obtain prisoners for interrogation. He wanted names and locations of Viet Com leaders as soon as possible.

Then they went to the PRU Compound where Gilbert gave orders for the Squad that had been with him all morning to start their firing weapons training. He had Kim join them, and Gilbert had the PRU Chief issue her the same weapons as the rest of the Tiger Scouts. He also informed them that she was now his interpreter. Then he had each Squad lineup for weapons inspection and to fire their weapons for record. Thirty percent of the Tiger Scouts met the criteria, and he then had them line up to receive their bonus money. Gilbert announced that when the other Tiger Scouts qualified he would pay them the bonus. He then told everyone that he would also pay each Squad a Hundred Dong when all its members qualified. He then paid two Squads a Hundred Dong each because all their members were qualified.

The PRU Chief asked how he could get a bonus. Gilbert said, "When all Squads are qualified you will receive a two hundred Dong bonus." Then he said, "My Interpreter, PIC Interrogators, and I are a Squad, and when we qualify we will get a bonus. When we are all qualified our Squad's bonus will be spent on a party. Everyone was congratulating each other and already suggesting a hundred ways to spend the extra money.

Gilbert told the PRU Chief to pick one of the winning Squads to accompany him on a mission that evening. Kim was standing there looking like she had a problem when Gilbert asked, "Can I help you?" She answered, "Where am I going to sleep?" Gilbert laughed, "Leave the things you won't need in my room. You will be going with us. We won't be returning for a few days, and I'll get you a room when we get back."

Gilbert had the PRU Chief drive them out to the Army Aviation Company where they were to get a helicopter ride to Kon Tum. Everyone in the aviation squadron thought they were going on a party but the Tiger Scouts were loaded for bear. None of the helicopter crew even guessed that Kim was a girl. She handled her weapons like a professional, and never spoke around them. They had about a half hour wait before they were to board the UH-1 helicopter, and Gilbert had everyone gather around him on the ramp where he informed them that this was actually going to be a snatch and grab raid but no one was to know until they had finished. The Squad just grinned and laughed. Anyone watching would have thought that they were having the time of their lives and maybe they were.

It was only about a twenty minute flight over beautiful country with small mountains on each side as they climbed. They all sat there with their M-16 muzzles pointed at the floor or deck of the helicopter. Gilbert had taught them to place the weapon on single fire and keep the muzzle down with their chin on the stock. It kept the helicopter's crew from getting nervous, and if a weapon accidentally fired it wouldn't hit any thing that would cause the helicopter to crash.

Gilbert suddenly remembered that he needed to give his Squad the locations were they could expect enemy fire to come from. He asked the crew chief if he could borrow a headset so he could speak to his Interpreter. The man located two headsets and handed him one and gave the other to the Tiger Scout he pointed out. He also adjusted the switches so they were on hot mikes and gave Gilbert a thumbs up signal. "Kim I want you to tell the men that the sappers (terrorists) are firing mortar rounds from the road and from the school yard," Gilbert said. "Do you want them to guard you until you are in the firebase?" She asked. "No, I want you to have them change into their civilian clothes now and leave their weapons except for the knives with us. We will cover them if they get into a firefight," Gilbert ordered. "Those Sappers (terrorists) will not surrender if they don't have weapons," Kim said. "I know. All I want is the mortar tubes and the shells. Leave the bodies," Gilbert answered.

There was no sound over the intercom. Except for the wop, wop, of the blades and the sound of the jet engine; it was very quiet until one of the gunners asked the pilot, "Sir, did you hear that?" "Are you still on the intercom Mr. Gilbert?" the Aircraft Commander asked. "Yes, just land to a hover on the log pad as close to the school yard as possible. As soon as my men have dropped off move over to the firebase entrance, and we will get out with all of our gear. It will be faster if your crew chief helps us unload but I'd suggest that your gunners stay alert. They are not to fire unless you are attacked." "Everyone hear that?" the Aircraft Commander asked. Then the same voice came on again and this time asked, "Is there really a woman aboard?" Gilbert laughed and answered, "Gentlemen, she is the hottest thing you've every carried. Be careful where you touch her when you help us unload." One of the pilots spoke, "I don't care what the Company Commander said, "I'm logging my time as combat."

When the helicopter came to a hover the PRU Tiger Scouts dropped off and disappeared into the crowd who were watching from the road. Then the helicopter hovered over to the entrance gate where Gilbert and Kim started unloading the weapons. The crew chief and two of the Army guards came over and helped them move everything inside the compound. The ARVIN Guards closed their side of the gate while Kim gave them hell for not moving fast enough. Gilbert asked one of the guards to notify the Battalion Commander that he had guests for dinner and cocktails.

The firebase Commander arrived ten minute later with half of his officers in tow and said, "It's really you. When they told me you had arrived I couldn't believe it." Gilbert answered him saying, "I came back to return the favor you did by letting me rest up for a few days." Then he introduced his interpreter and told him that he had a Squad who were working the streets so he wouldn't be woken up by a rocket or mortar attack. When Kim spoke and shook his hand the Lt/Colonel couldn't take his eyes off her. He kept checking and trying to see if she was some sort of mirage.

"Let's all go to the officers' club. It's time we get to buy you a beer," the Lt/Colonel said. "Kim you go and find us a place to bed down. I'll stay here and wait for our Tiger Scouts to return. They will need me to identify them or to back them up if they get in a firefight," Gilbert answered. Then he asked, "Colonel, can Kim stay with the Red Cross women for the next couple of days?" "I'm sure they won't mind." He answered.

It wouldn't be dark for an hour and Gilbert was hoping that his Tiger Scouts might get lucky and locate the enemy's mortar.

It was just about dark when he saw seven Vietnamese men approaching the gate. Gilbert recognized his PRU Tiger Scouts, and they had a prisoner who was carrying a mortar tube tied to his back. Three of the Tiger Scouts were each carrying a mortar round. They were laughing and waving their machetes in the air. One of the Army guards had phoned the officers' club to inform the firebase commander what was happening. Kim came running, arriving at the gate about the same time as the Tiger Scouts. Just behind her came the Battalion Commander and every officer who wasn't on duty.

Gilbert presented the mortar and three rounds of ammunition to the Battalion Commander. He then told Kim to take the PRU Tiger Scouts, and their prisoner to the barracks where they had been assigned. He also told the men through his interpreter that they would eat "C" rations in the barracks. After he had everyone on the way to their quarters with their weapons and gear he said, "Colonel, I'm ready for that beer." The Lt/Colonel asked, "Do you want us to lock up your prisoner?" Gilbert laughed answering, "I haven't arrested him yet. My Tiger Scouts will entertain him until I've had a beer, and then I will interrogate him. He may want to enlist in the PRU Tiger Scouts or just help us. What ever happens he will not bother you again."

While visiting the Battalion Commander over a beer in the Officers' Club Gilbert asked, "Can you notify the Load Master at the supply point on Qui Nhon that you have a platoon of ARVIN Soldiers needing transportation to your base tomorrow afternoon around 1300 hours?" Gilbert knew that Army CH-47 Chinook helicopters normally operated as individual freight haulers. They would arrive at a helipad on an airbase where a Load Master would direct them by radio from his jeep to a sling load or an internal load of troops and give the pilots the destination for the load as they were picking up. Gilbert wanted his PRU Tiger Scouts and two of the PIC Interrogators before dark. He planned on having them sweep the Viet Cong from the area that night.

He didn't tell the PRU Chief of the destination or where he was calling from. No one could know or even guess what he had in mind if the raid was going to work. Not even the helicopter's crew knew the destination until they were loaded. Even the Battalion Commander of the firebase at Kon Tun had no idea that Gilbert planned on breaking the back of the VC Communists Terrorist Cells and removing all the Viet Cong from the junction of the Ho Chi Minh Trail and Hwy 14 in one night. As Kon Tum was not in the same Province where the PRU Tiger Scouts were stationed the Province Officer In Charge (POIC) of either province wouldn't even hear about the action. Even the PRU Tiger Scouts wouldn't be able to tell anyone as they didn't know where they were or even cared.

The Chinook landed at 1400 hours, and Gilbert, Kim, the OD (Officer of the Day), the Firebase Commander, and his Tiger Squad were there to meet them. As they came off the rear ramp of the Chinook in their Black Pajama Uniforms they not only looked mean but acted as if they were going on a date with destiny. They all held up their M-16 in their right hands pumping them up and down in a salute as they passed by Gilbert. The ARVIN PRU Chief joined Gilbert,

Kim, and the Battalion Commander who asked, "Will you need any assistance?" He didn't seem too happy with the idea that they might be coming inside the Compound or that they were staying the night.

"No assistance needed," Gilbert answered. Then he said, "We will camp in the school yard and leave in the morning on the first Chinook heading back to Qui Nhon. If any of the men need water I'd appreciate it if we could stop your incoming water truck in the morning and use some of its water. If I have any wounded they will need water and possible emergency treatment. We will take our dead, wounded, and any prisoners with us." The Firebase Commander looked relieved.

Then Gilbert told the Lt/Colonel that they had located the girls that came in on the water truck yesterday. He had interrogated them and the driver. The women were working farm girls from 18 to 20 years old. One was married to an ARVIN soldier who had been drafted and had no income. Kim had insisted that they were VC so Gilbert had really checked out the water truck owner who was also the driver. Each of the girls paid him to ride in the water tank but that was all. He actually only provided the transportation.

The girls only wore silk pants with no tops like before, and the old man told Gilbert that he still helped lift each one up to where the ladder began, and assured him that they couldn't be carrying anything. The driver had purchased the truck from a Special Forces Sergeant who had rented it to him for two years before he went home. The man had bought and paid for the truck. The girls just paid for transportation, but the driver also sold Vietnamese Dong to the soldiers so they could pay the girls. He explained that it was a free enterprise business, and he was purchasing Bank CDs from the Ft Rucker Federal Saving and Loan Bank in the USA so he could move there when the war was over.

Gilbert told the Firebase Commander that if he wanted to protect the Firebase that he should get rid of the Colonel's Crapper as they had found documents showing it and measurements to the helipad, fuel dump, and barracks. The Lt/Colonel asked, "Could you take one of my officers with the Tiger Scouts tonight? He can keep in touch with us by radio, and if you need reinforcements I can send in one of my line Companies. If this happens I'll put my ARVIN Company on the Green Line." Gilbert answered saying, "I can use your officer. Send him over so I can brief him and let my Tigers smell him." He was joking about the smelling but the Battalion Commander took him serious. The 1st Lieutenant reported to Gilbert an hour later.

When the Lieutenant was introduced to Kim she just sort of snorted, and said something in French that wasn't very nice. Gilbert heard her and answered her in the same language saying, "If you don't behave. I'll assign the Lieutenant to you." She just looked startled and said, "You speak French." It wasn't a question, and Gilbert went back to briefing the Lieutenant as he showed him the men who were getting ready for the battle with the Dark Force. Their prisoner had accepted Gilbert's offer of money for the names of the Viet Cong in Kon Tum and had agreed to show them a factory making grenades. He also agreed to take them to where a nine tube multi rocket launcher was being stored.

The Leader of all the Viet Cong in this area was the headmaster of the school. The rocket launcher was hidden in a shed attached to the school room, and the rockets were under the classroom floor. This was one reason Gilbert had the PRU Tiger Scouts make their base in the school yard. He planned on hitting it first and cutting off the head of the snake. While they waited for the Dark Force to take over, the PRU Tiger Scouts put out targets at the edge of the school grounds facing a small mountain to the north and started practicing firing their weapons. Three more Tiger Scouts qualified with one of them making his Squad qualify for the team bonus. The most excitement came when Kim qualified but the two PIC Interrogators didn't make it.

Everyone lined up so Gilbert could pay and congratulated them. The Lieutenant asked, "Do they do this every time they are getting ready for a mission.?" Gilbert answered, "Not every time. These men are Viet Cong who are fighting the Communists. They know that they will not survive if the Americans leave Vietnam. It's a case of getting even before they are killed." Then the Lieutenant asked, "How about the girl?" Gilbert again answered, "She is Buddhist who is a Communist and a hard core VC. Kim was a Sapper (terrorist) Cell leader, and we caught her." The Lieutenant asked another question, "She's armed and sure knows how to shoot. Aren't you afraid?" "She has taken a job as my interpreter and really believes that the Viet Cong will kill me. She has only one problem and that's her love for money. I'm the only one who pays when a job is finished, and they have learned to trust me. Besides I don't sleep with her." Gilbert answered.

Tonight they were going to be in Viet Cong territory, and they had a new Lieutenant with them. It was just before the Dark Force took over when they grabbed the Headmaster of the school and found the rockets with its launcher. Gilbert had the Lieutenant contact the firebase, and asked for an Infantry Squad to pick it up. They took everything into the firebase and gave it to the Artillery Battery. The next target of the PRU Tiger Scouts was the factory that was manufacturing hand grenades for the VC.

Gilbert briefed the Lieutenant that this was to be a snatch and grab attack as he was sure the factory would be defended. The Lieutenant was to bypass the factory and leave it alone as Gilbert only wanted Viet Cong prisoners to interrogate. Gilbert didn't want to get in a battle for any building or ground. His PRU Tiger Scouts had been trained to get in and get out. They knew nothing about fighting for property or ground or even how to hold them. The factory acted like a magnet, and the Lieutenant went directly to it as a moth to a flame. The Viet Cong heard them coming and prepared booby traps. The VC had put grenades on a work bench with short fuses, and had the pins wired to the bench so if they were picked up they would explode. The Lieutenant lead the Tiger Squad into the factory and before Gilbert or Kim could get into the building to stop them someone picked up one of the grenades and it went off. The man's hand disappeared, but the grenade also set off other explosions. Another Tiger Scout lost an eye, and the Lieutenant took a blast in the chest.

Gilbert knew he should have listened to Kim but now they were in the factory building, and the fight was on to see who was going to get out again. Gilbert broke into one room and found himself with a Chinese AK-47 up against his head. He froze but there was nothing he could do. He was being hit by red hot empty M-16 casings, and the sound was one continues blast. It was Kim who had shoved her M-16 under his armpit, and hit the VC with so much lead in his head it had just disappeared. The enemy didn't get one round off. Gilbert was covered in blood but with not a scratch. When he could breath again he ordered, "Burn the building. Every one out." Then he turned to Kim and said, "Good shooting. Thanks." He spoke to her as a comrade and equal but in the excitement he had been giving his orders in Vietnamese. She asked, "Do I still have the job as your interpreter?"

CHAPTER EIGHT

The PRU Tiger Scouts in their black Pajama Uniforms were like dark shadows as they ran through the light from a burning grenade factory building in Kon Tum. As the flames climbed into the night's sky the people hid in their houses. The Viet Cong Communists were now the ones who were afraid. It was almost uncanny as there was no sound but the crackle of the burning buildings, and the small sound of explosions as the flames reached hidden grenades and weapons.

The Headmaster of the School, the Viet Cong leader had been killed. The Communist Tax Collector, Communist Enforcer, Sapper (Terrorist) Leader, and in actual fact every Communist Cell leader in the village was arrested by the PRU Tiger Scouts and taken prisoner. Everyone in the village knew when they were captured because soon after the arrest their houses would join with the flames of the grenade factory to light the village. This caused the Viet Cong to believe that the PRU Tiger Scouts knew all their names so they would break and run trying to escape from the village. The PRU Tiger Scouts would shoot all who fled. Those that surrendered were taken prisoner and held at the school. The Province Interrogation Center Interrogators were busy sorting them and identifying those who were the Communist Cell leaders. Each prisoner was told that if he identified another VC his house would not be burned, and he would be released if the names provided were not already on the Tiger Scouts VC list.

The Viet Cong started to turn themselves in while promising to show the PRU Tiger Scouts where the ammunition and explosives were hidden. By 0200 hours they had seventeen Viet Cong Officers and Cell leaders as prisoners. Many others had tried to escape capture and were killed. Two of the PRU Tiger Scouts were killed when they ran into an ambush set up by Communist Viet Cong. By daylight Gilbert knew that he had two men killed, two wounded, and one US Amy Lieutenant killed in the action (KIA). Besides capturing Communists VC the PRU Tiger Scouts had blown up a VC ammunition storage dump and found weapons hidden in houses throughout the village.

At 0500 hours the Battalion Commander sent one of his line company's field kitchen and their Cooks to the school yard to prepare breakfast for the PRU Tiger Scouts. After all the Tiger Scouts had eaten Gilbert let the prisoners be fed. While everyone was eating he sat down with his PIC Interrogators and Kim at a field table to interview the prisoners brought before him. His plan was to only take the hardcore Communists back for interrogation and to release the other prisoners. Gilbert's mission had changed from just seeking intelligence to buying time for the US Army to get out of Vietnam when the Paris Peace Accord was signed.

The families of the prisoners and people of the village had started to gather in the school yard to watch. The PRU Tiger Scouts were lying around eating and watching the proceedings while they waited for the Army CH-47 Chinook helicopter to arrive. Gilbert would thank each prisoner for helping them capture the Communist leaders. He then had them sign a paper making them a reserve PRU member, and paid them twenty-five Dong. What they didn't know was that he was going to forward their names as paid Province Recon Unit informers to Special Operations and the Tactical Operations Command (TOC) in Saigon. They would send the list to the American Embassy who would forward it to the negotiators at the Paris Peace Talks to trade for information on American Prisoners held by the North Vietnamese. When the North Vietnamese tried to bring their Army through Kon Tum, and after killing everyone on the list, no one was going to be helping them.

The Chinook helicopter arrived, and as they lifted off Gilbert looked down for his last view of Kon Tum and its firebase. He noticed that the Colonel's Crapper had disappeared and could no longer be used as an aiming point for the enemy mortars. An hour later they were unloading at Qui Nhon, and a USAF ambulance was waiting to pick up his two wounded Tiger Scouts. Gilbert had the two bodies of the Tiger Scouts who were killed placed in the pickup, and he personally drove it to the PRU Compound to deliver them to their waiting families. The body of the Army Lieutenant was picked up by a Dust Off helicopter and flown to the mortuary pad in Saigon.

While all the men rested in their quarters or went home to their families Gilbert' day had just started. First he had to order two wooden caskets. Then he had to chink the seams of the casket as a showing of esteem respect. He provided a Vietnamese Flag, the Vietnamese Cross of Gallantry, plus pay to each family for their loss. Then he had the names of the PRU Tiger scouts killed in action attached to the list of names of the new PRU reserve members, and delivered it to the Province Officer In Charge (POIC) knowing that it would be turned over to the North Vietnamese at the Paris Peace Talks.

When he turned the list of names for the PRU killed in action (KIA) to the POIC he just glanced at Gilbert and said, "You look like hell. Your job is to train and keep the PRU busy. You are not supposed to be partying all night." Gilbert answered, "We have seventeen prisoners in the Province Interrogation Center (PIC), and I've been helping train the Interrogators." The POIC answered, "Coddling prisoners is a waste of time. Get some rest and then come in to see me. I'll find something you can do."

The man hadn't heard a word Gilbert said and had no idea of what was going on in his own province. Gilbert was too tired to care. He didn't even wait for the mess hall to open but went to his room to sleep. When he walked into the room he saw that Kim was in the bed and snoring. He had forgotten that she hadn't been assigned a place to sleep. He pulled out the Army air mattress he had in his back pack, and after blowing it up he went to sleep while falling down upon it.

The next morning he woke up stiff and sore with the sun shining in the room's only window. Kim had gotten up and left but her bags were neatly stacked on the bed. Even his dreams were about the Province Interrogation Center (PIC), and how he could reduce the work load for the Interrogators while they found out the names of the Viet Cong leaders in Qui Nhon.

As he walked out of the Advisors Quarters he saw Kim talking to the PRU Chief. He walked over and asked if they had eaten yet. She answered, "I've been waiting for you." Gilbert told the PRU Chief to have the men clean all their weapons and prepare for another mission. He said all of this in English and waited until Kim interpreted the order. Neither Kim nor the PRU Chief acted as if anything had changed.

In a firefight everything changes. Some men become cowards, some change religion, some become heroes, and some become traitors but they all change. The night before when Gilbert had started giving orders during the heat of the battle in Vietnamese so his Tiger Scouts would understand they accepted this as a gift from Buddha, and today no one acted surprised that he needed an interpreter.

The PRU Chief said that he had spoken to his wife, and as they lived within the PRU Compound that Kim could live with them. Gilbert smiled and said, "That's perfect. She snores." If looks could kill Gilbert would have been knocked dead on the spot.

They drove into town in the PRU pickup truck and ate at a restaurant that Kim picked out. It was a large Vietnamese restaurant serving the local people. Kim was dressed in a traditional Vietnamese woman's dress with a hat to keep from showing her man's haircut, and Gilbert was dressed in his white shirt with dark pants wearing a shoulder holster with his 9mm automatic pistol. As they walked in the waiters greeted her by name, and Gilbert could see that he was causing some of the guests to get nervous. It wasn't the pistol he was wearing but the fact that he was an American with a known Viet Cong member. Even the manager wasn't too happy to meet him but smiled when Gilbert told Kim that this was a great restaurant for a new visitor to Vietnam. Then the manager made his first big mistake of the day. He thought that Kim was setting up Gilbert for some VC trap or using him to help the Communists. Gilbert had already decided that this restaurant was going to be the PRU Tiger Scouts next target.

When they returned back to the PRU Compound Gilbert had Kim take all of her things to the ARVIN PRU Chief's house to get settled in. She didn't know that it was only going to be for one night. While she was getting acquainted with the PRU Chief's wife and moving in with the family Gilbert took the PRU Chief outside. He told him that he had a Viet Cong area he wanted the PRU Tiger Scouts to sweep. He was to make a snatch and grab raid tomorrow night and take the prisoners directly to the Province Interrogation Center.

Gilbert then told him that he would be leaving on another assignment, and he was taking Kim with him. No one would replace them. Also that he was now in command of the PRU Tiger Scouts. Gilbert promised that he would be told the location of the snatch and grab raid for tomorrow night just before he left. They were to be taken to the airport and dropped off at the Army Company who had Slicks (UH-1) for a ride on one of their helicopters. The PRU Chief just nodded. Gilbert was so security conscious that he never told in advance his exact plans. They were leaving but not by Army helicopter but with Air America. Sometimes Gilbert wasn't sure himself until the last moment. It doesn't take many ambushes before a person learns to be unpredictable.

That afternoon all the Tiger Scouts who had not yet passed the sharp shooters test successfully qualified for their bonuses. Gilbert paid all the Squad bonuses plus the individual twenty-five Dong Bonuses before he told the PRU Chief that they would now go to the National Police Compound so he could test the PIC Interrogators. All the Tiger Scouts were standing around waiting for something. They weren't at all excited as Gilbert had expected them to be.

Then Gilbert understood and laughing said, "Kim, tell them they will all have to judge my score. I will shoot now." Someone put up a new target, and Gilbert borrowed an M-16 from one of the Tiger Scouts. Then he stepped up to the firing line. The excitement started building as everyone wanted to see the target after he had completed firing the test. Finally after everyone had seen and talked about it, they all started telling Kim their findings at the same time. Finally the PRU Chief got everyone to stop talking. Kim was grinning and said, "It is the finding of the Tiger Scouts that you have qualified and may receive your bonus." She hadn't finished before they were all shouting while pumping their M-16 up and down in a salute.

Gilbert drove while Kim and the PRU Chief rode in the front seat of the pickup with him. After they arrived at the PIC building Gilbert paid the PRU Chief a two hundred Dong bonus while saying, "You have done a good job, and your warriors are some of the best I've every fought with. I am going to give you some information that is for you and the PIC Interrogators only. This is not a secret but if used properly may save many lives." Then Gilbert had the interrogators meet with them under a tree away from the building, and where they could see anyone approaching.

First he told everyone that the Paris Peace Talks name had been changed to the Paris Peace Accord, and that the names of all PRU Tiger Scouts with the names of the PIC Interrogators were in the hands of the North Vietnamese Army. What he had told them was that they and their families were on the North Vietnamese death list. Next he told them that Qui Nhon city was a part of a planned North Vietnamese Army's route to Saigon. For their families to survive if this happened they would have to locate and destroy the Viet Cong organization before the Communists could take over.

He told the Province Interrogation Center (PIC) Interrogators that they had to do a quick interrogation of all prisoners and transfer those that were not hard core to the National Interrogation Center (NIC) in Saigon. The prisoners that were really dangerous were to be returned to the Chief of the PRU Tiger Scouts, and he would decide what to do. Then he informed them that every VC who was not an immediate threat should be given a chance to join the PRU Reserves and be released. Also that a list of the individual Viet Cong who had signed to become reserve members of the PRU would be furnished to the Province Officer In Charge of Qui Nhon and to the ARVIN Special Operations in Saigon.

Then Gilbert waited while everyone thought about what he had told them. The PRU Chief asked, "How will we be paid?" Gilbert told them that as everyman's name was replaced with the name of a Viet Cong who had signed the list they would use this persons name when being paid. If a man was killed his name would be sent to the POIC and Special Operations in Saigon so his family's name would disappear from the death list. They would have to capture enough VC to replace every man's name before their families would be safe. The Tiger Scouts would also have to make sure no Viet Cong Cell leaders or Officers survived who knew the VC's whose names they were using. The PRU Chief nodded that he understood and asked? "Where will you be?"

Gilbert answered, "I'm going to another Province who has the same problem and try to save them if the United States leaves Vietnam." Then Kim asked, "I am a Viet Cong. Why are you telling me?" Gilbert told her that she was a Montagnard and a Buddhist. She had already proven herself as a PRU Tiger Scout in combat, and all the Tiger Scouts were Viet Cong who were fighting the Communists. "Then he told her that the reason her hair was cut, and she had to wear men's clothing plus dark sunglasses was so she could disappear if Gilbert was killed. Then he asked her, "Do you want to leave now?" "No", she answered.

That evening Gilbert went to the USAF Officers' Club for dinner and a few beers. As Gilbert was eating he observed a man with curly bright red hair at the bar watching him. The man had no rank showing but was wearing Navy construction clothing. Gilbert thought to himself that this man looked like a scrounger. He guessed that he must be wanting something very bad to invade an Air Force Officers'

Club. Gilbert nodded towards him, and then invited him to his table. Gilbert said, "You must really need something big to bring you to a USAF Officers' Club. What do you need?" The Seabee answered, "We need some transportation and maybe some radio's if you got any." Now Gilbert knew this was his last night at Qui Nhon so he said, "How about a jeep. It doesn't run but you can have it if you take it tonight." The Seabee Scrounger just nodded. Then Gilbert said, "Come with me. Let's go for a ride."

Gilbert took him over to the CIA Warehouse and using their flashlights looked around telling him to pick out what he wanted and said, "Show me what your Seabees need. Anything that you pick out must be taken tonight. The man acted as if he just hit the jackpot. Gilbert had him sign a hand receipt for everything they moved out of the building. Just as they were getting ready to leave that Red Headed Scrounger saw a 50 Caliber Machine Gun and asked if they could have it. Not only did the Seabees get the weapon but Gilbert insisted that they take enough ammunition to start a war.

Gilbert asked him what they needed the machinegun for. The Scrounger answered, "We are getting sniper fire from the village near where we are repairing a bridge. Maybe the machinegun will scare them off. They've already wounded four of our men." "Here is something that should solve that problem," Gilbert said. Then he gave him a case of flare pistols and flares for them. "The next time a sniper fires on your men have them return the fire with these. They will cause the houses to burn that are being used as cover for the snipers."

The next morning as he walked into the PRU Chief's office for the last time; he saw a young woman with three small children (1 to 5 yrs old) sitting in the dust on the ground beside the door. Kim and the PRU Chief greeted him, and Gilbert asked who the woman was. Kim answered, "She is the wife of one of the PRU Tiger Scouts who was killed in a snatch and grab raid just before you arrived. She has been beaten, and the money she received when her husband was killed was taken by his father. After he took the money, she was thrown out into the street." Then the PRU Chief said, "Her husband's father is a Viet Cong, and we've left him alone because his son was one of us. She has no money, and I've let her sleep outside the PRU office inside the compound."

Gilbert said, "Send someone to the compound's mess hall for food, and see that they are fed. Be sure to check that her husband's name is listed as killed in action (KIA), and include it in the list you are sending to the POIC." Then he said, "Tell me the amount of money that was taken from her. I'll replace the money, but you must find someplace out of this city for her to live." The PRU Chief told Gilbert that her family lived in Saigon, and they would take her in." Gilbert answered, "Good, I will make arrangements for them to be flown to Saigon by Air America tomorrow morning."

Gilbert had one more bit of advice for the PRU Chief. "Don't forget that this man must know some of you if not all by sight. You'd better pick him up in one of your sweeps after his daughter-in-law has left the area, and before he has time to alert the VC in Saigon." "I'll do better than that. I'll have his name put on the list as a PRU member," the PRU Chief said.

Gilbert had done all he could for Colby's plan to delay the North Vietnamese Army in Qui Nhon. He had built a trained group of fighters and had shown them how to stop the Viet Cong. Best of all they had been tested in battle. Now it was time to leave, and he was worried that Colby didn't know where he was or where he was going next. He had for the moment forgotten that Bill Colby had started out with the OSS and was now of Ambassador Rank in the CIA. Bill had been monitoring the cash withdrawals from Gilbert's special accounts, and the reports of the State Department's requests from Paris asking where Gilbert was assigned. The Communists were insisting that his actions were interfering with the Peace Talks. With the State Department reports on where he wasn't and the finance sections reports of where the money was being spent, Colby knew where Gilbert had been. Colby also knew which areas he had told Gilbert to work which made it easy for him to follow his movements. In the meantime Gilbert was trying to come up with a plan to alert Colby where he was headed without letting the Special Operations Section or Tactical Operations Control (TOC) know.

When he signed out on leave at the POIC office only the Secretary was there, and she was the wife of the PRU Tiger Scout who had borrowed the key to the CIA warehouse for him. He asked the PRU Chief to drop him and Kim off at the air terminal building for charter airlines and Air America. While he made arrangements for himself and Kim to be put on the manifest to Plei Ku he also made reservations for the family of the slain PRU Tiger Scout to leave for Saigon the next day.

His plan was simple. Gilbert would request that he and Kim be placed on the manifest of an Air America flight to Plei Ku (USAF spelling Pleiku). He was sure that Colby would be told that he had been found, and where he was headed. The first battle for Plei Ku actually took place in 1965 but now Gilbert was going to try and set up delays to slow down the North Vietnamese Army on its march to Saigon when the Americans left. At the same time he wanted to save the PRU Tiger Scouts and their families. He wasn't even sure that he could do it. Gilbert only knew one thing for sure; if he did it and lived no one would care or even know it had happened.

The Air American plane was a single engine Porter aircraft, and it would take about a half hour flight before they would arrive at the USAF Base on Plei Ku. Gilbert was dressed in his CIA Advisor clothes of white shirts and dark pants with a shoulder hostler. He also carried his small automatic Browning in his left pants pocket. Kim was dressed in the Black Pajamas Uniform of a PRU Tiger Scout with dark glasses and the wide brimmed hat. She was carrying an M-16 with the CIA thirty round clip. As they approached Plei Ku the plane was flying over miles of forest. As Gilbert looked down he suddenly noticed something strange about the trees. They were all of the same height and ran in straight rows. He had been looking at them for almost five minutes before it dawned upon him that it wasn't a forest they were flying over but a French rubber plantation.

He had flown over a rubber plantation once in the III Corps area but nothing like this. This one ran from the edge of the airfield west to the Cambodian border as far as the eye could see. Plei Ku was a large USAF Base about three miles from the city with USAF fighter bombers, ARVIN AF fighter bombers, Australian Army helicopters, US Army and Marine helicopters all mixed up into one large military facility.

After they landed and were inside the base terminal Gilbert used the phone at the security desk to phone the Senior Province Officer. He was a man from the State Department who was trying to keep peace between all the military organizations in the area who were trying to defend their turfs from each other. It was a thankless job, and it took a professional trained man in diplomacy to make it work. Colby had phoned and alerted him that Gilbert was arriving to take over the PRU Tiger Scouts and the Province Interrogation Center (PIC). He had also informed the POIC that Gilbert was coming, and had the PRU Special Forces Sergeant Advisor rotated back to the United States. Gilbert hadn't been forgotten, and Colby had cleared the path. He had also left a message for Gilbert to come to his office in Saigon on Thursday the week after next.

They had to wait about thirty minutes before an unmarked jeep pulled up to the terminal entrance driven by a PRU Tiger Scout wearing the Black Pajama Uniform. Gilbert and Kim had been waiting outside by the exit for their ride with their barrack bags beside them. During the flight Kim had her face plastered up against the plane's window when they were flying over the Plateau du Kon Tum that was covered with a blanket of green grass. When the plane flew over the hills and forests she just looked away becoming very quite. In fact she hadn't spoken since they had arrived except to answer a few questions. Gilbert was worried as this wasn't like her at all.

As the jeep pulled up and stopped Kim suddenly came alert. When the driver came over to pick up Gilbert's bag she had already thrown her bag in the back of the jeep. Gilbert could see that she was watching the man intensely, and when she spoke to him it was in a Vietnamese dialect that he couldn't understand. The man answered her and Gilbert caught the word "Yes". Kim started talking so fast that even the PRU Tiger Scout who was driving didn't have time to do more than nod now and then. Gilbert finely hollered stop. "Not you," He said to the driver. "Kim, slow down and speak so I can under stand you. Who is this man?" Gilbert asked. She told him that he was from her tribe and that these PRU Tiger Scouts were Montagnard from the Sixth Tribe who had joined to fight the Chinese and Vietnamese Communists. The driver knew her family, and she just wanted to know about her friends. Gilbert laughed and suggested that she let the driver answer or at least say something. "By the time they drove into the PRU Compound it had turned into a homecoming. The PRU Chief turned out to be related to her, and a friend of her family. He also spoke English and seemed a little awed by Kim. He told Gilbert that the Sixth Tribe had thought she had joined the enemy, and her name was never mentioned in tribal matters. Now they had found out that not only was she a great fighter but the right arm of a great warrior chief.

Gilbert wasn't sure about her being the right arm of a great warrior chief but she was a fighter.

They decided that she would stay with the PRU Chief and his family. He also informed Gilbert that he would be honored by the tribe in a feast fit for a great warrior. All this sounded great to Gilbert until he was invited to go with the men to hunt dinner while the women started preparing the meal. The hunt was in the forest and amongst the rubber trees, all in VC territory. Gilbert was looking for wild pigs but the realistic Montagnards were looking for monkeys or anything that would go into a pot. Hunting wasn't very good, and the Chief told Gilbert that there were too many Viet Cong living off the land. It was in the heat of the day, and they had come across a stream where they all stopped to rest.

Gilbert decided that this would be a good time to cool off so he striped off his uniform and waded out in the stream. Suddenly he heard the men start hollering and everyone came running into the water laughing with much splashing of water. He stood up shaking the water out of his eyes and turned to see what all the excitement was about. Coming straight toward him was the largest snake he had every seen. There were twelve Montagnard PRU Tiger Scouts already hanging on to the snake, and all he had time or could do was grab its head holding the jaws closed while wrapping his legs around the body of the snake. He wasn't sure who was going to eat who, and all he could do was to just hang on while the rest of the hunting team finally got a hold of the snake, and the fight was on. With twenty men laughing and holding on to a twisting angry snake it didn't have a chance.

While eighteen men and Gilbert held the snake two men cut down a small thirty foot tree and lashed the snake to the tree. Gilbert found himself stark naked and sweating up a storm but there was no way he was going back into the water. With all the noise and excitement of the battle with their dinner Gilbert was glad that no Viet Cong had heard them and showed up or he was afraid the Montagnard would have just added them to the pot. They offered him the honor of carrying the front of the snake but he had been within two inches of its nose far too long and gallantly said, "The honor was rightly the Tribe's Chief as he was only a guest. He said he would walk along just behind them so if it started to get away he could catch it." All the Montagnards laughed.

Just thinking of the meal ahead was making him wish that Colby would phone and tell him that he was needed in Saigon or on the moon. This tribute by the Montagnard Tribe wasn't going to be his finest hour.

There are all kinds of snakes in Vietnam. Tree Vipers, Cobras, Green Vipers, Bo constrictors, and even one called a One Step Snake. The Vietnamese called it that because if you got bitten you took one step and were dead. Gilbert knew all about the snakes that the Dark Force had put in and around the Garden of Eden, but he hadn't thought that he would have to eat one. What he didn't know was that he was going to be tested many ways before the Bright Force's accepted him as one of its mission teachers. Strange are the ways of man but stranger yet are the tests of the Bright Force.

That evening the Montagnard Sixth Tribal council met in the PRU Compound at Plei Ku. A small fire was lighted to see by, and all the warriors gathered around with the Tribe and PRU Chief, three Sub-Chiefs, and Gilbert in the center ring. There were a hundred warriors squatting in circles around the center ring with about twenty of the PRU wives preparing food and carving up the snake. Two men brought in a 2½ foot tall Clay Jug and placed it in font of Gilbert and the PRU Chief. Then they were each given a three foot long piece of bamboo made into giant straws. They placed a bamboo 10" by 10" cross over the mouth of the jug. The jug was full of dark red wine. One end of the cross was inserted into the wine about five inches. They called it a stick. The PRU Chief showed Gilbert that a warrior had to suck up and drink the wine until it dropped below the end of the stick.

To become a member of the Sixth Tribe a warrior had to drink six sticks of wine. The snake saved Gilbert for as he drank each stick he would think of the meal to come and have to leave the circle to throw up. After drinking six sticks of wine he was named a member of the Sixth Tribe and presented a brass bracelet with the marks of the sixth tribe. It was made to fit on the arm and was not a full closed circle.

By this time everyone was drinking more sticks of wine and dipping in to the giant pot for boiled snake and vegetables. One of the women was using a large wooden ladle to fill the wooden bowels that the PRU Tiger Scouts were bringing up to the cooking pot. There were also over three dozen children all wanting some of the meat. When Gilbert had his bowl filled he suddenly had to throw up again and handed it to one of the children. He laid his back against one of the trees to watch the party, and that's all he remembered. After the party was over the women who had a husband would go through the Tiger Scouts and find their man. They would carry, drag, or lead him home if he could still walk.

The next morning Gilbert woke up in his quarters with a hang over that hurt so bad he could hardly see. Kim had found him and dragged him to his quarter's feet first. His head felt like it had hit and bounced against every rock in the compound. His clothes were washed and folded on a chair with his bracelet on top of his shirt. His jump boots were cleaned and shined. All the PRU Tiger Scouts wore the US Army stile of combat boots with canvas sides except Gilbert and Kim. He could only find jump boot that were small enough for her. She seemed to think that they were special because they were like those he wore.

He needed water and grabbed his canteen to drink. After drinking trying to put the fire out in his belly he suddenly had to run outside to heave again. After twelve aspirin and more water he finely staggered over to his office. Gilbert thought to himself that no one could have ever have had a hangover this bad before. He sat at his desk, put his wastepaper basket under it, then placed his forehead on the desktop, and prayed that the Lord would take him from his misery. As he sat with his forehead lying on the desk top he heard Kim come in. He knew it was her because she whispered asking, "Are you awake?" He didn't answer, and a few minutes later he opened his eyes, and he was looking down into a pair of doe slanted beautiful eyes looking up at him. When he hadn't answered, she had come over, and got down under the desk so she could look up to see if he was awake. All Gilbert could think of was to say, "Boo". Kim said, "There is radio call from the POIC. He says he has to talk to you."

Gilbert took the radio phone she handed him, "This is Gilbert over." The POIC told him that this was an emergency. There is an Air American plane that crash landed on the Plateau du Kon Tum northeast of Plei Ku. He was to take twenty PRU Tiger Scouts and place a guard around the plane and not let anyone near it. The load was classified, and another plane was being prepared to pick up the load. Two UH-1 Slicks would pick his team up at the helipad at his compound in ten minutes.

Gilbert came alive, and told Kim to get twenty Tiger Scouts, and that they should prepare for two days in the field. "Tell the PRU Chief to meet me at the Log Pad." As he left the office he picked up his M-16 and an extra clip. The Tiger Scouts had come running. By the time Gilbert arrived the PRU Chief had already picket twenty men and placed them so there would be ten in each squad. "You all right?" He asked. Gilbert answered, "I've been better. Were these men at the party last night? They look great. Why haven't you got a hangover?" Just then the helicopters came in, and the men started loading before they even touched down. "Kim get in that helicopter," Gilbert ordered as he climbed into the second one.

As soon as he got aboard he pointed to his ears, and the crew chief handed him a set of earphones. He asked the pilot if he could talk to his team in the other helicopter. While he was waiting he said, "The other helicopter has one of my people who knows the area we have to go into. Let them land first, and please circle so I can get a good look at what we've got to defend." When Kim had ran to get in the helicopter that Gilbert had pointed toward he had reached down and picked up her nap sack. It was twice as heavy as he expected. Now he had time to ask a few questions that he should have done before except for the headache. Somehow it had disappeared.

"Kim, can you hear me?" Gilbert asked. When she answered one of the pilots turned and looked at Gilbert but didn't say a thing. "Be sure that no one goes near the plane. Keep the area clear so another plane can land to pick up the cargo. Have the men recon by fire. What's in your kit bag?" Kim answered. "I put in some gifts for my mother." Someday Gilbert would learn not to ask what a woman carries in her overnight bag. The Aircraft Commander asked Gilbert over the intercom if he still wanted the other helicopter to lead the way. Gilbert answered, "She knows the land, and this is her peoples' hunting ground. If she says don't land I'll call in the Air Force and blow everything up." The pilot said, "It looks like someone is walking around and waving." Gilbert spoke up and said, "Same rules apply. Only if she says not to land make a stop and pick up the person walking around but make him run to us out of the range of small arms fire from the wrecked plane." The Aircraft Commander said, "I like the way you think."

Everything went like clockwork and after Kim's team had landed, the helicopter Gilbert was in circled the area. Gilbert was looking to see if any VC were hiding in the brush waiting for them. It looked as if no one had found the wreckage yet. The pilot had tried to land on a small dirt trail but the plane must have been so heavy that when it hit both landing gear had been torn off. The plane was on its belly with one wing tip into the ground. The pilot's door was off, and he was the one walking around waving at them.

Just before they started the approach the pilot said to call when they were ready to leave as there was a radio relay aircraft airborne that would send them the message. Gilbert told him that they would move as soon as the load was picked up to a point at least five miles from the accident site. "Do not land here again even with a smoke signal as it won't be us," Gilbert said. "I like the way you think," the pilot answered again.

As soon as they were on the ground Gilbert pointed toward the direction the plane had came from and sent his squad to cover the approach. He watched as his men moved out and stopped along but off the path. Then he turned and checked the other direction where Kim had her team. He had a hand held radio that was on her frequency. "Give the radio to your team leader," Gilbert ordered. She was looking at him, and he held up his right arm above his head as he gave the order. It was a hand signal for her to come to him. She advanced the radio one frequency, waited until she got to him, and advanced Gilberts radio to the same frequency so the two squads were again able to talk to each other.

Gilbert walked over to the pilot who was standing guard, and introduced himself. At the same time he looked into the Air America Plane. There was one man in the co-pilots seat who had been crushed by the boxes that the plane was carrying. It looked as if he was killed when the load shifted. One of the boxes was broken open and his blood was soaking the contents of large Dong Bank notes. The plane was loaded with boxes of Vietnamese Dong. No wander they didn't want anyone to see what was in the plane. Gilbert noticed that there were also a couple of the crates that had killed the co-pilot lying outside the crushed right side of the plane which hadn't broken open. Gilbert spoke to the pilot in French asking, "Do you know why the engine stopped?" The man didn't understand and asked him to speak English as he didn't understand French.

Gilbert told Kim to only speak to him in French as the pilot didn't understand the language. Then he pointed down the runway where he had his team take cover, and then turned pointing in the direction where the other team was in hiding beside the trail. Then they both squatted and pointed at the ground while drawing a map with a stick. While they were doing this Gilbert had been telling Kim that he wanted her to have two of the men to move one of the crates to a position ten feet from the aircraft and hide it in the grass. He told her that the Dong (money) was going to be worthless in a month or two from now but today it could buy a lot of things the tribe needed. Kim just nodded and trotted off towards the men she was going to use. Gilbert couldn't help but noticed that she ran like a girl and was worried that the Air America Pilot would notice. The pilot was so worried about the money that he wouldn't have noticed if she had been running with no shirt on.

A few minutes later a Caribou aircraft with no marking circled them and set up a base leg for landing. It landed on the trail and came to a stop just ahead of the crashed plane. The pilots didn't get out or even shut down the engines. Four American men came out of the back and lowered the ramp. Two men went over and looked at the plane. Then they started to take off the engine cowlings. These men were professionals who knew this type of aircraft and without hesitation started cutting the engine mounts away from the firewall of the plane. The other two men started cutting or disconnecting the wings. In thirty minutes they were connecting the cargo wrench to the fuselage of the wrecked plane, and had begun pulling it into the cargo hold. Kim had returned from her mission, and just nodded at Gilbert.

In less than an hour with not even a thank you the Air America plane was taxiing to take off. The co-pilot waved in answer to Gilbert's wave as they left with everything including a dead co-pilot still in his seat but minus the wings, engine, prop, landing gear, and one case of money. The case was too heavy for one man to carry, and there was no way to put it in a helicopter without questions. He had all the men open their back packs, and two men broke the case open while Kim told them that this was the tribes' money. This was not a problem for the Montagnards as the tribe owned everything. Gilbert divided the money so it fit into the twenty back packs they had with them. Then they threw the broken box in amongst the other wreckage that had been left.

Gilbert was getting nervous as this had put them in the area for almost two hours. There still was no indication of any North Vietnamese soldiers or Viet Cong but he knew that they had to be on the way. One of the PRU Tiger Scouts took the point and started off as if he knew where they were going. After an hour of a fast march they came upon a small stream where they turned and headed due east. Gilbert thought that this would be a good place to call for a pickup but before he could get Kim's attention the pace was picked up until everyone was trotting. Gilbert knew that when twenty PRU Tiger Scouts didn't stop to drink or fill their canteens, and then started to really move out that this wasn't a time to question them on why. Something was just ahead or just behind them. Gilbert wasn't sure what it was.

After fifteen minutes the pace picked up again. Now Gilbert was really worried. He had run five miles each day before joining the CIA, and he wasn't sure if he could keep up. Just being a member of the Montagnard Sixth Tribe wasn't enough. He must be getting old. Suddenly they came up over a small rise in the ground, and he could see the men ahead of him. They all broke into a flat out run, and he heard shouts like children calling to each other. By the time Gilbert got to where he could see what was ahead of them he was the last man in line. Everyone had passed him and some were already out of sight.

Suddenly he could see, and there ahead stretching out before him was a Montagnard Camp. The outer ring was cooking fires and what appeared as pits or food storage areas. Then he saw a ring of small camp sites where the families lived on the grassy plain. They had spears with the butt end down and with the points pointing up in front of their straw sleeping mats. Some of the spears had metal tips, and some had burnt sharpened points but all were gathered together and held by one of the tribe's brass bracelets.

The next ring of campsites for families had platforms built out of wood and branches that was two feet above the ground. There was no shelter on the platform which was about six by six feet. These sites had Springfield rifles, M-1, and some times a M-16 held muzzle up with one of tribe bracelets holding them together. Then he noticed that while they didn't have shelters they were of different sizes. Then came platforms about 10 by 10 feet, and their weapons were covered by a poncho or something to keep them dry. Next came a smaller circle of platforms that held a thatched shelter built on them that would keep the occupants dry. Gilbert had been walking trying to follow Kim as she moved toward the center of the village. She stopped at a platform in the ring next to the center platform that was at least 20 by 20 feet with a lodge built on it which had a thatched roof and screened sides that were rolled up.

This was the home of the PRU Chief who was also the Chief of the Sixth Montagnard Tribe. Kim's family was one of the sub-chiefs, and part of the inner circle of the governing council. No wonder everyone was excited. One of the PRU Chief's sons came and asked him to share their home. Gilbert unslung his M-16 with the CIA Clip and placed it with the others in front of the platform. When he placed his bracelet around the muzzle to hold it with the others he heard gasps of astonishment. The word spread and children came from everywhere to see the strange round eyed member of the tribe. Gilbert was asked to eat with the PRU Chief's family but he was practically out on his feet, and made excuses that he needed to rest. All he could think of was that snake and fighting to see who was going to eat who. He drank water from his canteen. Then he laid down just inside of the mosquito netting on a sleeping mat and fell asleep.

The next morning he woke up to a beautiful day with a light breeze blowing. The breeze brought him the clean smell of grass and flowers, and cleared the camp of mosquitoes. The green grass kind of shimmered in the breeze and looked as if it was waves moving across a pound. The children were all sleeping, and the camp was quite. Gilbert used his radio and gave their position as being next to a Montagnard Camp on the Plateau du Kon Tum near a small stream ten miles northeast of the accident site. Two Slick Helicopters would make the pickup within an hour.

This time it was Kim who looked as if she hadn't got any sleep but when Gilbert told her that they were to be picked up in an hour she started rounding up the PRU Tiger Scouts. When everyone was assembled Gilbert counted them and there were twenty PRU Scouts but he didn't recognize a few of them. He asked Kim if these were the same men that they had left with and she said, "No, five of the men decided to stay home. These men had taken their place. The PRU Chief would understand." Gilbert wondered if he would when he found out about the money they were carrying.

After returning to the PRU Compound Gilbert made out his after action report for the CIA Province Officer In Charge (POIC), and then went to see his PRU Chief. He told him what the load was, and told him that somehow a case of the Vietnamese Dong Bank notes had gotten left behind. The PRU Chief thanked him and told Gilbert that he had two of his sub-chiefs out purchasing items for the village so it would be spent before it became worthless. Kim had been standing listening and said, "I think Gilbert is concerned because five of the Tiger Scouts didn't return with us?"

Gilbert then went on to explain that the North Vietnamese had a list of all the names of the PRU members, and that they planned on killing them if the Americans left. The PRU Chief laughed and said, "Our agreement with the American Government is to furnish one hundred warriors to fight the North Vietnamese Communists. This we do and these men change all the time. It does not matter what the North Vietnamese do. We fight them as a tribe. They do not keep our names but just kill all Montagnard what ever their name is. What can we do that will help you?"

"You can come with me to the CIA warehouse on the airbase. Bring ten men and the 2&½ Ton Truck and I will give you enough ammunition for the tribe to last a year." Gilbert said. Then he added, "I would like to sweep all the Viet Cong areas around Plei Ku." "That's easy. The VC and North Vietnamese bases are in the rubber plantation to the north and east of the airbase. You were there with us when we hunted for meat for our fest." the PRU Chief answered and then he went on, "The land is part of our tribe's hunting ground but the POIC has told us that we are not to go there as it is part of the US Marine's territory. He also said that if we damage a tree or marked one that the United States would have to pay the French plantation owner a thousand dollars."

After thinking about this for a moment Gilbert told him that he would inform the US Marine Commander that the PRU Tiger Scouts would be hunting on their land. He just wouldn't tell him what they would be hunting. Then the PRU Chief got into his jeep and directed Kim to follow them in the truck with the ten Tiger Scouts that she had picked to go with them. They headed for the airbase and the CIA Warehouse. After they went inside Gilbert told them that he wanted the Tiger Scouts to take all the M-16 with the CIA clips, and to see that everyone in the tribe had one. Then he told Kim that they should take everything the tribe could use to fight with, and that the PRU Chief was going with him to visit the Marine Commander. They would be back before the truck was loaded. "Let's see if we can find the Marine Commanders office?" Gilbert told The PRU Chief.

They drove toward the runway looking for a sign or something that designated the US Marine headquarters. They came across two Air Police in their jeep waiting at a cross street. Gilbert asked them to show them the way. The AP led them about four blocks down the street they were on, turned left for a block, and there it was. It wasn't the Marine TOC but an office they used to coordinate their operations with the USAF. They were in luck as the Marine Major was on the airbase meeting with the USAF base operations officer and was in the office.

After waving thanks they got out and entered the building. Gilbert showed his CID ID, and they were escorted into the Commander's office. He was a Marine Major, and when Gilbert told him that they would be hunting in the rubber plantation on the plateau starting that night, he told them that it would be no problem. Gilbert said, "The hunt may take three days. If any VC or North Vietnamese attack us during our hunt we will hit them with everything we've got." The Marine Major answered, "I'd hope you do. If you need backup in case they attack radio me on our tactical frequency."

As they drove back toward the Warehouse the PRU Chief said, "I noticed that you didn't say anything about the rubber trees." Gilbert laughed and said, "Why remind him. The Marines are part of the US Navy, and they don't think about trees as being a problem on the ocean."

The PRU Chief said I am now speaking as the Sixth Tribe's Chief, "I am putting the woman you call "Kim" in charge of moving all the supplies and items we are purchasing to our camp on the plateau. She is related to me and the daughter of one of my sub-chiefs. Our tribe needs many strong leaders, and she can produce many fine babies." Gilbert felt proud and kind of like a teenager in heat. He thought that the Chief was going to ask him to make love to her. Instead the Chief continued, "She must not engage in any direct combat unless they are attacked. We have many young men who can give her strong children. Our tribe will need strong leaders no matter what happens. It's too bad you are too old." Gilbert was a little taken aback until he remembered that the life expectancy for Vietnam men was only forty years, and in the Chief's eyes he was already an old man.

Kim told the PRU Chief that they had found some C-4 explosive in the CIA storage building and asked if they could take it. When Gilbert wanted to know what they were going to blow up the PRU Chief explained that the women used it to blow up VC mortars. They had a tool that could remove the firing cap of a mortar round. When they located any Viet Cong hidden mortar rounds they would pull the caps and replace the propellant with C-4 which would cause the round to explode in the tube. The explosion was large enough that it destroyed the building and everything near the tube. They would change all the mortar rounds as the VC never figured out that it wasn't a defective mortar tube and used the rounds that didn't go off in the blast in another mortar. Gilbert added the C-4 to the list he was preparing to leave in the building showing that the items were used for the Province Recon Units.

After the truck was loaded Kim and ten PRU Tiger Scouts with the driver followed the PRU Chief and Gilbert north on Hwy 14 to a side road about five miles out of Plei Ku. They went about a mile up to the beginning of the rubber plantation. They stopped and hid most of the load in a side gully. As soon as this was completed Kim, and ten of the PRU Tiger Scouts loaded their individual back packs with a hundred pounds of the goods and started off through the rubber trees heading toward the grass lands of the plateau about a half mile to the east. Kim never looked back as she disappeared in the trees. It was the last time Gilbert saw her as she led ten strong young men each carrying their hundred pound load at a trot heading home.

The shadows of the night were drifting across Hwy 14 when Gilbert and his PRU Tiger Scouts silently joined the shadows among the rubber trees. The Bright Force had given up the jungle of trees to the Dark Force and forty PRU Tiger Scouts started their penetration of VC territory. This was no snatch and grab raid but a direct attack on the enemy's base of operations. Gilbert had gone from being an interrogator to a Sixth Tribe Sub-Chief leading forty of his Montagnard warriors on an attack against their enemy.

They were searching for marks on the trees that would indicate a trail to the enemy's lair. There it was a round paint spot at eye level height that was in line with another dot and as you followed the dots they led to small clearing between the lines of rubber trees. Gilbert had his Tiger Scouts in four columns about ten feet apart and ten feet behind each other. They were spaced so they could provide coverage from VC fire and still be looking for the entrance to the underground base.

The point man stopped and faded into the earth. Now they were searching the tops of the trees searching for the guards. Each column was a squad and Gilbert had given each squad leader a metal beetle or clacker. One clack came and then five minute later another clack. They had found two guards. The system was simple. The PRU Tiger Scout who had located an enemy soldier got to take him out. Gilbert took out his 25 Browning automatic and every one waited. One of the enemy guards came crashing down from the tree he had been hiding in. Gilbert stood up and using his small caliber automatic fired, the other guard fell out of his tree, and two Tiger Scouts were on him with their knives. A Tiger Scout that had been climbing up the tree trying to get to the guard but had to stop climbing when the other falling guard had alerted him to the danger. Gilbert had taken a chance that the sound of a small caliber gun wouldn't be loud enough to sound an alarm and it worked.

Every one played statue until even Gilbert was satisfied that no alarm had been sounded. Two of the PRU Tiger Scouts went over and started searching for an alarm wire, and one of the men held up his arm. They had found it. The Tiger Scouts slowly opened the trap door leading into a tunnel going down into the earth.

To crawl into the belly of the Dark Force was not Gilbert's idea of fun or what soldiers did on their night off. The PRU Tiger Scouts formed themselves into two squads. Then the first twenty men slung their M-16s over their left shoulder and with their long knives in the right hand silently slipped into the tunnel one by one. Then the second Squad waited for Gilbert to enter the darkness.

It was dark as pitch in that Viet Cong tunnel, and Gilbert held the knife in his right hand as he slowly eased himself through the trap door. In the tunnel he swept the knife back and forth trying to find his way in the darkness. It was as if he was trying to cut through blackness so thick that he could taste it. All Gilbert could think of was the stories of how the Viet Cong would sometimes catch poisonous snakes and tie them to hang down from the ceiling. It was gut wrenching work.

The first squad had disappeared into the darkness with out a sound. Now he was leading the second squad into the same darkness but all he could hear was the sound of his own breathing and the thump, thump of his heart. Gilbert was sure that the enemy could hear the thumps as they were so loud his ears rang with the noise. Suddenly he felt a hand touch his back.

CHAPTER NINE

It was a hot afternoon right after the lunch hour while the forty Montagnard warriors of tribe six were taking a brief siesta (resting) before heading into a French Rubber Plantation for the coming night's patrol when Gilbert wrote into his journal; "these men are like American children on Halloween day. Being forced to wait for night to come before going trick or treating makes them think that the sun is standing still."

When Gilbert and the Montagnards had visited the CIA warehouse on the airbase they had taken all the British made torches (flashlights), batteries, and backpacks still in the warehouse. Now there were forty packs lined up on the compound's drill field with a torch laying on the top of each one. Gilbert had been busy on the phone talking to the Commanding Officer of the US Marines who he had met that very day. He had asked for four Marine Slicks (UH-1 helicopters) to pickup his PRU Tiger Scouts at around four in the afternoon and to insert them at a spot of his chousing near the rubber plantation. His only problem was convincing the Marine Major that if the Marine helicopters inserted them he would know what part of the rubber plantation for the Marine patrols to avoid.

What the Major didn't like was the idea that Gilbert would not tell anyone where they were headed until after the helicopters were airborne. He finally agreed to the plan after being told that Gilbert was looking for an elephant trail leading into the plantation and didn't know the location as he hadn't found it yet. When Gilbert explained that the

PRU Tiger Scouts would be establishing an ambush site that night, and would not have to be picked up by helicopters until the next day. He agreed to the plan.

The PRU Chief squatted next to Gilbert as he sat on the office steps looking across to where the Tiger Scouts were resting in the shade from the trees at the edge of the compound. He could see that the men all had the new automatic M-16 rifles with the CIA 30 round clips. They were all dressed in the black pajama uniform with the soft brimmed hat. Each man also carried a long knife strapped to his belt. Gilbert removed his shoulder holster and handed it to the PRU Chief, "Take care of this. I'll need it when we return." Gilbert still had his 25 cal Browning automatic in his left pant's pocket, but anyone observing them from a distance wouldn't be able to tell one soldier from another. When he was on patrol or in the field Gilbert's protection came by looking the same as the other PRU Tiger Scouts. His teachers at the CIA Blue U had taught him that a CIA Agent should blend in and not become a target. For two and a half years now Gilbert had survived and was living proof that the CIA Instructors were right.

Gilbert had counted the backpacks twice, and there was still one for each of them including him. He had gone over, lifted one of the packs, and it weighted at least a hundred pounds. Montagnards must have something against backpacks that weighted less than a hundred pounds. Gilbert had made sure that there were "C" rations for two days, and a thousand round of M-16 ammunition in each pack but had left the rest of the load up to the PRU Chief.

He hoped that they weren't planning on running up to their home camp to say hello after the night's patrol. These men liked to ride in helicopters and trucks but they thought a ten mile run was a just jaunt across the compound. If Gilbert had looked at himself in a mirror he would have seen a man, 40 years old, hard as nails, who's only problem was that the men he was running with were all twenty or more years younger he was. They watched him continually in amazement as if they were wondering how an old man could make them work so hard. There was no way they would fail to stay up with him, but they were young men with no fear and loads of energy. Tonight they were going to test him again by letting him lead them into an underground Viet Cong maze of tunnels.

What Gilbert didn't know was that they had been there before and knew that it was a North Vietnamese Battalion Surgical Hospital. It had been built for a North Vietnamese Infantry Division that was going to attack Plei Ku when the Americans tried to leave Vietnam. The hospital was fully equipped and supplied but not manned. It was guarded by a platoon of Viet Cong Guards who lived in a few huts near the rubber plantation located along an elephant trail while waiting for the signing of the Paris Peace Accord, and the arrival of the hospital staff.

At 1500 hours Gilbert had the PRU Tiger Scouts loaded into two trucks with their backpacks for the short ride to the USAF Base where they waited for the Marine helicopters to pick them up. The Marines arrived with four Slicks (UH-1) and three gun ships. Gilbert climbed into the first Slick (UH-1) helicopter with his interpreter, and took a seat near the left waist gunner where he could see the area they were flying over. Gilbert asked for earphones and mike so he could talk to the Aircraft Commander. The helicopters didn't take the time to shut down their engines or stop the blades from turning. When the PRU Tiger Scouts and their gear were aboard the helicopters they lifted off in a trail formation with the Cobra gun ships in a position of one on each side, and one behind where it could protect the entire formation.

Gilbert directed the Aircraft Commander to fly towards the northwest and stay to the west edge of the rubber trees. He told the Marine helicopter crew that he was looking for an elephant trail leading from Cambodia to the east, and entering the rubber trees from the northwest of the French Rubber Plantation headquarters. Gilbert had never been there, but he had been briefed by the Montagnard Chief about the location of a Viet Cong platoon of irregulars who were guarding a North Vietnamese facility.

They were now flying north of Plei Ku with Hwy 14 off to their left about one click (kilometer) when one of the gunners called out that he could see a small stream and a trail running southeast away from the highway headed toward the rubber trees. Gilbert also saw four huts near the road and stream. This had to be the area he was looking for. "Pilot continue flying north until we have passed the buildings and hold 2000 feet altitude. Hold this heading for another mile, and then turn east letting down to five hundred feet, then turn back to the southeast, and fly until we come to the edge of the rubber trees. When we come to where a trail goes into the rubber trees land, and let us off."

Then Gilbert said, "when you takeoff stay low and continue south until you are back to Plei Ku, City."

There were no elephants to be seen but their dung was everywhere. The stream was just a trickle, and the elephants had made it into a large pound or mud hole. No water actually entered the plantation. Gilbert spend the time while the PRU Tiger Scouts off loaded the helicopters sketching the location of the stream and where he had seen the buildings. His plan was to contact the USAF and ask them destroy the area with a B-52 mission from Guam. The Air Force liked these kinds of targets. No reporters to observe, write stories, or show pictures of crying children, and damaged schoolhouses. They didn't think about or care that all the elephants that would be killed or a Viet Cong platoon, and their families who would be vaporized. Gilbert knew that the elephants were being used to carry rockets and supplies for the Viet Cong. He wasn't a hunter of animals, but he could tell from the dung and tracks that there were no small elephants or mothers with nursing babies in the herd using the water hole, which meant that these were animals used by the Viet Cong as pack animals. Colby had told Gilbert to do everything possible to slowdown the North Vietnamese Army to give the Americans time to evacuate Saigon after the Paris Peace Accord was signed. One way was to destroy the North Vietnamese means of transportation for military supplies. Gilbert still didn't know what the Montagnard PRU Tiger Scouts had in store for him.

The helicopters had disappeared when the Montagnards shouldered their backpacks, and the point men had moved into the trees leaving the rest of the Tiger Scouts to follow them. There were seven backpacks still lying on the plain when Gilbert moved to pick one up. As he was lifting it into position with a grunt and groan as the pickets who had been guarding the LZ (landing zone) came in and shouldered the other six packs. They fell in behind Gilbert and followed him as he joined in behind the last squad of PRU Tiger Scouts as they entered the rubber trees. Gilbert noticed that the Montagnards were all grinning as they followed him into the forest of trees acting as if they were going to a party. It was strange being amongst forty armed fierce looking young men moving into the dark shadow as night was falling, and not a sound could be heard.

While silently moving amongst the trees for about fifteen minutes the column of PRU Tiger Scouts suddenly stopped. The soldier at the end of the squad that Gilbert was behind turned around, came back, touched Gilbert on the arm signaling him forward. Not a sound was made. He was now at the head of the column with the point men. They had found the VC Guards and the entrance to the underground tunnels.

After the fist group of twenty PRU Tiger Scouts had entered the tunnel it was Gilbert's turn to take the lead for the next group going into the darkness of a Viet Cong tunnel. He found himself trying to cut the darkness with his long knife. Just before Gilbert entered the blackness he noticed that the Montagnards who had followed him were still grinning from ear to ear. The last two men at the end of his column were dragging the two dead VC guards besides carrying their 100 lb backpacks but they didn't seem to mind the extra work.

It was so dark Gilbert thought he could taste the blackness. There was no sound but he could see a bright flash now and then just ahead as the men used their torches. Suddenly the flashes disappeared. The tunnel had widened out, and Gilbert moved over to his right until he could touch the wall. He was using his knife to feel for the wall, and at the same time he noticed that the tunnel had become large enough where he could stand upright. He could touch the top of the passageway only by holding his knife up above his head. Suddenly he felt a hand touch his back. It was one of his PRU Tiger Scouts asking for his backpack. This is one time Gilbert didn't insist on carrying his share but gladly slide out of it. Gilbert turned back and flashed his torch for a second. All of the Montagnards were squatting resting and waiting for Gilbert to give instructions. With that brief flash of light he saw that the tunnel was splitting into two passageways with the left one heading down to a lower level. He knelt and felt the floor with his hand. It wasn't dirt but felt like bamboo, and his knife acted as if the right wall was of wood. He reaches out with his hand and felt wood. The wall seemed to be made up of boxes stacked along the passageway from the floor to higher than he was tall.

Gilbert found it hard to think about anything but being below the ground in VC territory. The darkness seemed to be squeezing him making it hard to breath. As a CIA Interrogator he was used to working and dealing with people under stress. Suddenly he realized that his PRU Tiger Scouts had to have been here before. Being an Interrogator he knew how men reacted to stress and this wasn't the way. He still didn't like being in the dark even if these deadly kids were having fun. Gilbert started to think clearly and fast.

If the Montagnards had been here before they knew that there was no danger from the darkness, and that they were just having a little friendly fun at his expense. He flashed his torch on and saw that there was a small corridor or tunnel leading off to his right and slowly climbing upwards. Other torches had now come on, and the men were heading down the tunnel. He now saw that the boxes were actually empty coffins, and that the tunnel had widened out into what appeared to be a medical operating room with four tables. He also saw that the two PRU Tiger Scouts who had been dragging the two dead bodies had pulled the pins on two hand grenades and were placing them under the bodies of the two guards. If the bodies were moved they would explode. He didn't have time to more than glance at what the PRU Tiger Scouts were doing as his interpreter told him that they only had forty-five minutes before someone would be checking on the two guards.

All the Tiger Scouts were opening their pack packs and removing C-4 explosives, fuses, and timers that they had gotten from the CIA warehouse. They were working fast and ignoring Gilbert. Someone had opened his backpack and had removed the C-4 explosives that he had been carrying. Gilbert changed back into what he had always been an interrogator and seeker of information. He started to look for the office of the hospital commander. There was this corridor next to where he was standing that must lead to something important so turning down the tunnel he followed the corridor to its end. His interpreter called out, "watch out for booby traps. We have to leave in thirty minutes." Gilbert looked at his watch and moved down the side tunnel. He soon came to a door. It was unlocked but it was also the first door he had seen since he had been underground. He opened it, and with the beam of his torch he was able to see two diesel powered generators. They still had the yellow operating tags and the American Clasp Hands stenciled on them. They were new Caterpillars mobile power units painted yellow. He also saw two large fuel tanks and banks of batteries.

Gilbert couldn't read Vietnamese but at home in the United States he was an engineer and mechanic who could recognize that the batteries were for an emergency lighting system for the hospital. He went over and studded the control panel for a moment and then threw the switch that turned on lights throughout the complex. Gilbert then went back to the operating room where he found a pair of rubber surgical gloves and two hand grenades. Returning to the fuel tanks he pulled the pins, and placing a grenade in each glove dropped one into each tank. When the rubber softened from the fuel there was going to be a big explosion and someone was going to be surprised.

Upon returning to the tunnel leading back to the surface he saw that twenty backpacks were lined up, filled with drugs, and medical supplies. He knew then that half of the PRU Tiger Scouts were going to be headed back to the Montagnard Camp. Gilbert found it strange that these PRU Tiger Scouts didn't rely on him for anything. They knew the mission and accepted the fact that all Vietnamese hated them. Gilbert was a member of the tribe, and he provided them with the things their people needed. They also accepted the fact that every thing they found that belonged to the enemy was his, and what he didn't want belonged to the tribe. One of the PRU Tiger Scouts pointed to the far end of the operating room and indicated that it was what he was looking for. He hurried over, and there was another door. When he opened it he was looking into the Chief Surgeon's office and hospital records' center.

When the lights first came on Gilbert saw that the ceiling was of split bamboo as was the floor. The walls seemed to be of concrete and everything was painted white. The floor in the operating room was of concrete but when he checked the walls with his knife they were found to be made of stucco over dirt. While inspecting the power generator room he saw that it was actually near the surface and the roof was designed to be removed. This hospital was laid out and designed to support a complete North Vietnamese Division. Once the Paris Peace Accord was signed, and the killing began there would be no need for secrecy, and the roof of the power room would be removed.

There was no time left so Gilbert stuffed his backpack with all the papers he could find in the office, and then ran to catch up with his squad who were starting to move out. No one bothered to turn out the lights as this North Vietnamese facility was no longer an asset to the enemy. Gilbert looked at his watch, and they had only been underground for twenty-five minutes, but with his claustrophobe it seemed like hours.

As they exited the tunnel complex the men that were headed to the Montagnard Camp left the rubber plantation near the elephant's trail but split off toward the northeast and the plateau du Kon Tum. Gilbert and his group headed deeper into the rubber trees moving fast. Gilbert was beginning to feel the pace as his back had never fully recovered from the injures received when he was ambushed at Tuy Hoa. The pain pills were wearing off, and he didn't have anymore with him. He had no pain when he wasn't moving and just hadn't realized that he would be running ten miles while carrying a 100 lb backpack. Gilbert thought to himself that Colby couldn't know that he had sent him up against a North Vietnamese Infantry Division with just a hundred crazy Montagnards.

Gilbert wasn't worried about the twenty men who were headed home as this was also going to be his last patrol before reporting back to Bill Colby in Saigon. They had walked for hours through the rubber trees, and just before dawn the PRU Tiger Scouts stopped. Everyone rested and drank from their canteens while opening up their "C" rations. No one smoked, and there was no talking. Gilbert's interpreter came and motioned for him to follow him as they crawled to where Gilbert could look toward a clearing. He was looking at a dirt runway almost 2000 feet long. There was a Canadian Beaver aircraft in French colors parked at the far end of the runway near a small hanger. The plane had a high wing was powered by a single radial Pratt & Whitney engine of 650 HP that would carry six passengers plus their luggage or if the seats were removed a thousand pounds of freight. The Canadians used the Beaver on floats or wheels but this plane was on its wheels with a tail wheel for steering while taxing. A little further to the right he could see an old French Mansion in the just before dawn's light.

North Vietnamese soldiers were starting to stroll around the grounds doing what all men do before breakfast and the day's labor. The North Vietnamese Infantry Division of 1500 men was coming to life. Gilbert was looking at the headquarters of a complete enemy Infantry Division, and its supporting Units. They had taken over the French Rubber Plantation knowing that the Americans were under orders not to damage a tree or to do anything to disturb the Plantation. Gilbert crawled away where he could use his radio and knowing that no one would believe him, or would answer a call for an air strike, he would have to improvise. Instead of requesting an air strike Gilbert called for a radio relay connection to Bill Colby. He had to talk to Bill and knowing the mind set of the Jewelers in not letting a Contract CIA Agent talk directly to the head of all CIA Operations in Vietnam. Gilbert asked for a direct link by informing the L-19 from 'E" Battery 82 Artillery 1st Cavalry Division that Lt/General Moriggia wanted to speak to Bill before he ordered breakfast. It's the kind of request that Generals make now and then, and no one questions this kind of message.

Within five minutes Bill came on and asked, "What can I do for you General?" Gilbert told him that he had read in the Army Times that there were a lot of North Vietnamese supplies being moved over elephant trails, and then gave him a location east of Hwy 14, and north of the rubber plantation near Plei Ku of some elephant trails. He suggested that a B-52 bombing mission starting within a mile of the rubber trees would not disturb the Plantation owners and should destroy the trails. Then he went on and said that he would remain in the area until after the bombing raid to observe the results. Colby answered saying, "the Army Times sure covers a lot of subjects. Stop in and see me when you are in Saigon." Now all they had to do was plant the CIA charges of C-4 where they would do the most good and move out.

After he signed off Gilbert felt a hand on his arm, and he rolled over to see what his interpreter wanted. He had disappeared, and the PRU Montagnard Chief was holding out a bottle of his pain pills and a pair of binoculars. They were laying side by side about 3000 feet from the French Mansion and the hanger in what appeared to be a vacant field. The PRU Chief had a piece of camouflage netting that he pulled over them. He was also dragging a rife case that he placed between them.

Speaking softly Gilbert asked, "How long have you been here?" "We entered the plantation right after you got into the North Vietnamese Hospital. We've set C-4 charges in the fuel dump, ammunition dump, latrines, and on the buildings including the main house. They are all set to explode tomorrow morning when the latrines are in use. I wish we could be here to see the excitement," the PRU Chief answered.

"I've been waiting for you to arrive. I figured you would need these pills," the PRU Chief said. "How do you plan on getting the Tiger Scouts out of here before the charges go off?" Gilbert asked. "The Vietnamese have patrols and guards to keep people out not in. All of our people are inside the perimeter of the camp or hidden out in this field between the runway and the ammunition dump. No one moves around until after dark and before midnight. We will leave tonight when the enemy patrols move out. Let's go back to your observation post there is something you must see," the PRU Chief answered.

The PRU Chief and Gilbert crawled back to the observation post, and Gilbert didn't even offer to help as the PRU dragged the rifle case with them. Gilbert planned on resting while they waited for night to fall before attempting to leave the French Plantation. It was going to be another five mile run to get out of the rubber trees, and he hurt just thinking about it.

As Gilbert observed the enemy camp he saw that a dirt road ran around behind the small hanger, and the east end of the runway heading toward Cambodia. Along the south side of the runway just within the edge of the rubber trees facing the runway were five Chinese built tanks with 90mm guns, an aviation fuel dump, and further down the runway were the elephant leg irons but no elephants. Toward the west end of the runway was the ammunition dump, and on the opposite side of the runway were the tents of the Infantry Division. All were in line with the runway and under the rubber trees. Strung above the equipment and tents between the trees was camouflage netting. Any aircraft or helicopter flying over would only see the French Mansion, hanger, and a French aircraft. Then Gilbert spotted the Tiger Cage just inside the tree line near the hanger and saw that there was someone chained to one of the cages. He couldn't identify who it was but only that it wasn't a tiger.

The PRU Chief said, "Take a look through the glasses at the airplane and the barrels of fuel near the hanger." As Gilbert studied them he saw that two men were refueling the Beaver and seemed to be getting it ready for a flight. As Gilbert watched the PRU Chief told how he was watching when the plane landed the day they had penetrated the plantation, and a North Vietnamese General with three Colonels had came in on the plane. The passengers had been met by another General, who seemed to be the Division Commander, and a Frenchman with a large white hat who he guessed was the French Plantation owner.

As Gilbert watched one of the men who had been helping to refuel the plane got in and started it. Gilbert recognized that it was the daily preflight check. As this was going on the PRU Chief said, "This must have been a really big planning session. With all the fuel they are putting in the plane it must be going to Hanoi. I've brought my special built German snipers rifle just in case you think that this target has priority. All the charges are set to go off just before daylight tomorrow. There is no way we can change them. Should we attack before they leave?"

Gilbert looked at the wind sock, and it was hanging about quarter down so that it indicated a 10 MPH wind. It was oblivious that the plane would have to taxi down to their end of the runway before taking off. "Let me see your sniper rifle?" Gilbert asked. It was a single shot bolt action 50 cal rifle with a rugged scope and two triggers. It was designed to fire with no drop of the projectile, one trigger to be used when the target was lined up, and the other to fire the weapon when it was touched.

The ten miles per hour wind would carry all the sound away from the headquarters building and from the Infantry compound towards the forest behind them. If only one shot was fired and it hit the pilot there was a good chance that no one would hear it or guess what had happened. Gilbert handed the weapon back to the PRU Chief and asked, "Are you a trained sniper?" "I was trained in the United States by your Army," he answered. "I'll act as your spotter," Gilbert answered.

Then he told the PRU Chief that the target would be the left pilot's door of the Beaver as it was taking off. After the plane was in position for take off it would have 25% of its flaps extended. In this position the shooter wouldn't be able to see the target. The engine would be at full power as this was a short runway, and the plane was heavily loaded. They were in a good position as they were five hundred feet from the end of the runway and about 200 feet to the side away from the camp. The plane would be airborne, 10 feet in the air, flaps at 15 degrees for climb and power reduced for climb when the plane would be in position for the shot. The target would then become visible to the shooter. The range would be about 1200 feet; the angle up would be about 10 degrees, with a 5 degrees angle off the target. The shot would be into the wind, and with a 50 cal steel projectile there would be no drop. There would be only time for that one shot, and if it was successful the enemy would never knew it had been fired. Gilbert said, "I'll call off the position of the target and what it is going to be doing. Your target is a spot two inches forward of the door handle and half way up from the bottom of the door."

Gilbert noticed that the PRU Chief had put up a tripod to help steady the rifle and had it was aimed approximately in the correct direction. The camouflage netting covered both the spotter and shooter with just the end of the scope on the barrel of the rife clear of the netting. Gilbert turned his binoculars to watch the house as the plane was taxiing to pickup its passengers. They had come out of the house and stood watching the plane. There was the pilot, who was a surprise to Gilbert because it was a woman in a white jump suite. She had short dark hair and was shorter than the men who were talking to her. Gilbert wondered to himself if it was the same French woman pilot that Gentle Ben had told him about when he first arrived in Vietnam She had radioed for everyone to stop shooting while she landed her Beaver at a rubber plantation in the III Corps area. Here she was talking to two Vietnamese Generals, three Colonels, a Captain and a Frenchman in a large white hat. The plane only carried five passengers plus the pilot so Gilbert guessed that one of the North Vietnamese Generals, the Captain, and the Frenchman were seeing the others off.

The pilot walked around the plane for a quick preflight, and then she climbed into the pilot's position through the target door. The others entered via the cargo door, and as they taxied out one Vietnamese General, Captain, and the Frenchman in the white hat waved good by. Gilbert glanced at his watch, and it read 0900 hours on a bright sunshiny day. The passenger's bags had been loaded by the man who taxied the plane from the hanger, and he spoke briefly to the pilot before starting to walk back to the hanger.

"The target is moving toward the takeoff position. You should be hearing the sound of the engine as the pilot is checking the magnetos. Flaps are down to takeoff position. The plane has started to move. The tail wheel is off the ground. The plane's wheels have cleared the runway. Climb Power, Flaps to Climb, ten feet off the ground, half way down the runway, coming into your kill zone." Gilbert called out as the plane took off.

Suddenly the gun fired. Gilbert didn't take the binoculars from his eyes but continually watched the Beaver as one wing started dropping, its nose dipped, then rose steeply above the skyline, the wing leveled off, and then the nose dropped again. There was no way the plane could now clear the trees at the end of the runway. Suddenly it nosed down until it hit the tree line 500 feet from the end of the takeoff runway and still flying ten feet in the air with full throttle. The plane hit the trees and exploded into a ball of flame.

Gilbert looked over at the PRU Chief, and saw him as he was putting the gun back in its carrying case. He hadn't even glanced at the plane until after the explosion. None of the PRU Tiger Scouts had moved. If he didn't know better he would have believed that he and the shooter were alone. The North Vietnamese soldiers had run toward the crash site trying to help control the flames. It appeared that they were trying to stop the smoke from the fire from attracting the attention of other aircraft. The enemy officers were getting everyone back under the trees, and someone in an old US Army jeep was pulling a drag made out of logs over the road covering up tracks. A crew of soldiers was shoveling dirt over the planes' wreckage while others were putting up camouflage netting above the burned wreck of the aircraft.

While resting under their camouflage netting the day slowly drifted by. Gilbert and the PRU Chief lay within 1500 feet of the North Vietnamese Infantry Division Headquarters, and Gilbert actually felt good about what was to happen. Suddenly he heard the tinkle of water, and it wasn't raining. He glanced over at the PRU Chief, and he was peeing as he lay on his side. "Sorry but I always have to pee after a kill, "he said.

The PRU Tiger Scouts were all hiding as if they were snipers waiting for a target. They were so close to the runway that the North Vietnamese never suspected that anyone let alone ninety PRU Tiger Scouts could be hiding in an open field of foot-tall grass. All the enemy patrols had moved out beyond them amongst the rubber trees on the alert for anyone coming to see what caused the smoke. It was actually safer then being in any other part of the Plantation as they waited for the night shadows to announce the time for their departure.

When darkness had claimed the plantation the PRU Chief sent a Squad to pick up the prisoner that Gilbert had seen chained to a tiger cage by the hanger. When the Squad returned they brought with them an ARVIN Soldier who was found locked in a tiger cage. He had been a prisoner of the North Vietnamese for two years, and everyone including his family thought he was dead. The ARVIN soldier was wearing the pants and shirt of his guard who had been killed by his rescuers. The PRU Chief placed him with the Montagnard Squad who was assigned to protect Gilbert.

It was now 2000 hours, and the PRU Chief had the entire company of Tiger Scouts lined up in a column of twos with their faces camouflaged with the black charcoal from the burned aircraft site. Gilbert was towards the rear of the PRU Tiger Scout's column with the rescued ARVIN Soldier waiting beside the road leading out of the plantation.

Gilbert asked his interpreter if he knew what they were waiting for. He was told that they were waiting for a North Vietnamese Patrol to pass through the guards posted at the gate. Gilbert walked forward and joined the PRU Chief. The men were all squatting just off the road and waiting. A few minutes passed, and along came a Lieutenant with a patrol of twenty North Vietnamese soldiers in single file. He used a flashlight to show some papers to the guards and an argument started. The PRU Chief, Gilbert, and one Montagnard soldier carrying an AK-47 out in the open moved forward to hear what was going on. The Lieutenant was waving his arms and shoving a paper at the sergeant in charge of the gate. It seems that the password had been changed, and no one was allowed to leave the camp. The Lieutenant was waving his written orders, and at the same time shouting that he was to reinforce the hospital or something to that affect.

The Lieutenant was shaking his fist and threatening the guards with some dire promise. Gilbert walked up behind the Lieutenant along with the PRU Tiger Scout who was carrying the AK-47 and without saying a word the man pointed it at the guards while Gilbert pulled out his long knife shoving the point against the ribs of the Sergeant. The Sergeant stopped arguing, saluted, and stepped back from the Lieutenant and the knife. The Lieutenant just glanced at Gilbert and shoved his papers back into his carrying case while ordering, "Forward." The North Vietnamese soldiers just stood up and followed their officer. They didn't care one way of the other.

The PRU Chief signaled his Tiger Scouts to close ranks with the North Vietnamese soldiers so they appeared to be of one company. Gilbert and the Tiger Scout with the AK-47 just stayed in position watching the guards. Gilbert would holler out now and then, "No talking. Pick it up." At the same time he would hit a backpack of one of the Tiger Scouts with the flat of his hand so the guards could hear it. Every time the guards heard him hit the backpack of a man passing through the gate they would laugh. Gilbert knew that both the North and South Vietnamese Sergeants controlled or accented their orders by a hit with a fist or club. There were 20 North Vietnamese Infantry soldiers and 90 PRU Scouts that past through the gate, and Gilbert was afraid the guards would get suspicious if they didn't hear the sounds of blows being given.

Gilbert moved up beside the Lieutenant with the Tiger Scout who had stood with him at the gate. When they had moved out of sight and sound of the guards he and another Tiger Scout jumped the man while at the same time all the other Tiger Scouts attacked the North Vietnamese patrol. In less than three minutes the fight was over and not a shot was fired. The Montagnards took all the weapons, backpacks, and threw the bodies off the road amongst the rubber trees. They didn't want another patrol to discover them until morning.

It would be another four hours of fast marching before they were out of the rubber plantation, and already Gilbert felt the pain in his back. On the march the PRU Chief would order a halt for ten minutes of rest every hour. During the second break Gilbert used the radio to call the relay L-19 Army aircraft to request a helicopter pickup for eight Tiger Scouts at 0700 hours. He would pop two smoke grenades, one red and one blue. The pickup was to be at a point near the junction of an elephant trail, and the road leading toward Cambodia just out of the rubber plantation. The L-19 pilot came back and said that there was a B-52 Bomb warning "notam" posted, and that no flights were to be conducted in that area until after 0900 hours. Gilbert then changed the request to 0900 hours, and said that the call sign was to be Hotelman. He was hoping that someone would notify Colby that he had been located. Gilbert also asked that the ARVIN's Soldier's family be told about his rescue, and that he would be dropped off at the USAF Base Operations Building.

The PRU Chief had came back to visit Gilbert before he gave instructions for the helicopter pickup. He had told Gilbert that the Montagnard Camp was going to be moved, and that the PRU Tiger Scouts would be needed to help. They were going to stay on this road until after it crossed Hwy 14, and they were in Cambodia. Then they would turn north towards Kon Tum and their Camp. He also handed Gilbert a paper telling him that it was the list of names that the Province Recon Unit used for paying the Tiger Scouts. Give this to the Province Officer In Charge (POIC) and tell him that everyone was killed in the B-52 bombing raid. "What about the six men who are with me?" Gilbert asked.

"Have them dropped off at the firebase near Kon Tum when you are finished using them," the PRO Chief answered. "I'll send a years pay for each of the men on the list as KIA along with the Squad when they are flown to Kon Tum. How will you get paid for the Tribe' services in the future?" Gilbert asked. "The Vietnamese are our enemy. You have paid the tribe for a year of our services in advance. I'll give the men their money the same as I do now," He answered.

"I like the idea of telling the POIC that all the Tiger Scouts were killed in the B-52 bombing of a French Rubber Plantation. Everyone will be classifying everything secret about what has happened the past few days, and covering up a USAF mistake that never happened. They will all be occupied and working so fast that they wont fuss about paying the families KIA compensation. I'll bet that the North Vietnamese will never tell or admit that they lost a complete Infantry Division and two generals to a PRU Company of Montagnards. They may even claim it was a B-52 bombing raid that caused all the damage. How much further before we are out of the forest?" Gilbert asked. Before the PRU Chief could answer they heard the sound of an engine and the clank of tank tracks coming from behind them on the road. It sounded as if one of the Division Infantry support tanks was coming and everyone scattered to the sides of the road.

It was dark as pitch under the trees but the moon still gave a little light on the road. The tank was coming slowly up the trail a little faster than a man could walk. The Tiger Scouts had melted into the trees boarding the small dirt road. Gilbert could see that there were about ten Vietnamese soldiers hanging on the outside of the tank as it approached. The tank commander was standing up in his hatch calling down instructions to the driver helping him stay on the road. They must be headed to the beginning of the entrance road of the plantation to establish a guard post to prevent visitors from going to the French Mansion. The tank must have left the compound just after they had attacked the North Vietnamese patrol.

Gilbert noticed that the soldiers hanging on the tank were dressed in ARVIN Uniforms so he guessed this was to turn back any US Army units that might be interested in what had happened without raising any suspicions. These soldiers were not interested in anyone leaving the plantation so they just hollered insults at the Tiger Scouts as they passed. Both groups were dressed in the same uniforms except the PRU Tiger Scouts had their faces marked with black charcoal. As the tank started passing Gilbert; the Montagnard with the AK-47 rifle ran out and jumped up on the back of the tank. No one on the tank tried to stop him. All of the Tiger Scouts watched as he crawled up to the tank commander who was shouting down at the driver, and then without saying a word dropped in a hand grenade through the hatch. As the Montagnard soldier dropped it, he turned, and dived off the tank rolling off the road into the darkness under the trees.

Before he hit the ground other Tiger Scouts started firing their weapons, and then the grenade exploded. The tank stopped. It didn't burn and no one climbed out. Just to be sure someone threw another hand grenade in the tank. The Tiger Scouts got up from their rest brake, and began quick marching trying to get as far away as possible before another visit by a North Vietnamese patrol. As Gilbert walked pass the tank he looked to see what kind it was but all he could determine in the dark was that it was built in China, and the infantry soldiers lay where they had fallen.

They were marching toward the west, and Gilbert was beginning to see shadows on the ground from the trees. The sun was slowly creeping up behind them. The PRU Chief came back and joined Gilbert. "It's about time for the morning's big surprise. We've been out of the rubber trees for the last half hour. The elephant trail is just a few yards ahead. When you turn off on it go about a 1000 feet and wait there for your helicopter. We will continue west until we cross Hwy 14 before we stop to watch the B-52s drop their bombs." the PRU Chief and Headman of the Montagnard Tribe Six said to Gilbert as he shook his hand.

Gilbert lead his Squad about a thousand feet south through the scrub brush on the elephant trail until they came to a cleared area where a Slick UH-1 helicopter could touch down to pick them up. He radioed the helicopter telling them that they were ready. The sky had lightened, and there wasn't a cloud to be seen. The Army aviator answered telling him that the B-52s had started their bomb drop, and he had to hold his position until the ground stopped shaking. Suddenly the light faded from a dust cloud so thick that the sun looked like a blurred disk. The bombs were actually hitting a mile to the north of the Plantation, starting just east of highway 14. The USAF was right on target but suddenly the North Vietnamese Hospital they had mined with C-4 charges blew up with a small mushroom cloud joining the explosions of the Air Forces 500 pound bombs. The ammunition and fuel dumps of the North Vietnamese Infantry Division came next. Smoke and fires from all the building where they had set C-4 charges now started joining the explosions of the bombs. Everything started burning.

The ground heaved up and down causing Gilbert and his Squad of Tiger Scouts to fall but everyone except the ARVIN Soldier they had released from the tiger cage was laughing and rolling on the ground. The ARVIN Soldier was so scared he shit in his new pants and smelled so bad that the Montagnards even started laughing harder. Half of the rubber trees were destroyed or burning and individual North Vietnamese Soldiers were seen running out of the forest toward the highway.

Gilbert and his Montagnard soldiers took up a position just to the edge of the cleared area, and he set off the two smoke bombs for the helicopter. He radioed the helicopter, "two smokes, one red and one blue, over" "Roger, we have you in sight," the pilot replied. Just then Gilbert felt the ground tremble and heard an elephant trumpet. Everyone turned to face the sound, and through the smoke charging directly toward Gilbert came a bull elephant. He brought his M-16 up and fired a three round burst. The elephant didn't even slow down but issued a scream that made his hair stand up. In that split second Gilbert saw that the elephant was carrying a large load and his trunk was curled down between its tusks. There were four other charging elephants all carrying large loads. Gilbert fired another burst with the

M-16 with no affect. "Aim for the loads. Fire on automatic," Gilbert shouted.

The charging Bull elephant was so close that there was no time to run or even to jump aside. Gilbert was firing his M16 almost straight up at the elephant's load. The ARVIN Soldier had thrown himself down at Gilbert's feet hiding his face in the dirt. Suddenly everything turned red as he felt himself flying backwards through the air trying to swim trough warm blood. His M-16 was wrenched from his hands never to be seen again. Then nothing.

CHAPTER TEN: GOING HOME

Gilbert began hearing the sounds of a helicopter's blades going, "Wop, Wop." He could not see anything but he could feel that he was lying on the deck of a flying helicopter, so they must have been picked up. He could move his arms and legs but when he tried to speak he gagged on blood, and it tasted awful. He put a hand up to one of his ear, and felt a large piece of meat covering it. He pulled it away and dropped it. Gilbert didn't hurt but he couldn't see, and he could hear someone asking him in English if he could help? He gagged again but managed to ask for water. Someone put a canteen in his hand and started to hold up his head to drink. He pushed the hand aside and poured the water on his face. He pulled off peaces of elephant skin that covered his noise and eyes. Gilbert could now see, and he slowly sat up with his back leaning against the pilot's console facing toward the aft of the helicopter. He was in an Army UH-1 Slick, and he needed more water. This time he washed his mouth out before pouring the rest over his head.

Something smelled bad, and it wasn't just himself. He was looking at six Montagnards soldiers watching him while grinning from ear to ear, and one was holding a three-foot piece of ivory tusk on his lap. The ARVIN Soldier was seated on the deck of the helicopter away from the Montagnard Tiger Scouts near the starboard gunner's open door. Without much effort Gilbert could tell that, he was the one who stank from crapping in his pants. It was the second time today, and he had not been cleaned up from the first time.

Gilbert grinned at the Tiger Scout who was carrying the ivory tusk, and he held it up so Gilbert could see that it was not complete but had been broken off in an explosion. It must have been one hell of an elephant hunt as everyone was covered with blood and guts of elephant. Gilbert waved to one of the gunners, and when he leaned over to hear better Gilbert told him that he wasn't hurt. In addition, that he wanted to go to the Army's Division Helipad. The gunner spoke on the intercom, and signaled to Gilbert with thumbs up.

During the dry season in Vietnam all, the Army helicopters after landing for the day and before shutting down the engine would have a water truck meet them. The crew would shoot water into the engine air intake for two minutes claiming it helped clean the inside of the engine. Gilbert didn't give a dam about the engine he just wanted the water truck to hose them down. The first man to get a hosing was going to be the ARVIN soldier and then himself.

The only thing that happened different from Gilbert's plan was that the crew chief insisted on hosing out the cabin area after the engine was shut down before he would let his passengers use any the water. The Army Helicopter Operation's Officer, a Captain, met the helicopter accompanied by two MP (military police) and stood upwind while he greeted Gilbert. "Welcome to Plei Ku Army Heliport. Do you need medical attention?" the Captain asked. He didn't offer to shake Gilbert's bloody hand.

"I'm CIA Contract Agent Joe Kelly. These Montagnards are my guards." Then pointing to the ARVIN Soldier he said, "This ARVIN has been a prisoner of the North Vietnamese for the past two years. We just rescued him last night. My men and I have not eaten for two days. The ARVIN has been starved for the past two years. I would like a shower for my men and some breakfast if that's possible?"

"Have you any ID?" the Captain asked. "I have a CIA walk on water card. Would you notify the Senior Advisor for the District that I'm here? Tell him that Gilbert H. Morriggia (CIA pseudonym) would like to report rescuing a prisoner being held in a B52 target area." Gilbert had added the bit about the B52 bombing, as he knew that this was a subject that even the President of the United States would be briefed on. He just needed Bill Colby to know what was going on. As the Montagnards were being lead to the showers Gilbert speaking in a low voice to the Captain said, "Have your MPs keep an eye on the ARVIN Soldier. Do not let him near any weapons. His mind is a little mixed up from the beatings he has been receiving by the North Vietnamese. Our weapons need a good cleaning. Can you find us a gallon of WD-40?" The Captain just nodded.

Gilbert joined the men in the shower room. After he had scrubbed and scrubbed, he finely gave up from trying to feel clean. He opened his backpack, brought out a razor, and using hand soap shaved. He had one set of green underwear, a white shirt, dark pants, belt, socks, and low shoes. He had lost his M-16 and his shoulder hostler with its pistol but he still had the small 25 cal Browning automatic pistol he carried in the left front packet of his pants. His wallet with 100 dollars in American Script, and $500 Vietnamese Dong was still in the backpack. He also had $5000 Dong stuffed into a dirty sock and another $500 dollars of US script in another a one. The MP's had found a dozen cleaning rags and a quart of WD-40 oil. Someone had given them an old Army blanket, and the Motagnards were all busy field stripping and cleaning their weapons while they squatted around the blanket. It made Gilbert feel as if he was part of their family.

Gilbert joined them and did the same thing to his 25 cal automatic pistol. The showers and cleaning of the weapons had taken about an hour. One of the MP (military police) told Gilbert that if they followed him he would take them to a mess hall for breakfast. The men had all washed the elephant blood and gore from their uniforms. They had let them dry in the sun while they cleaned themselves and their weapons. Gilbert just cleaned his jump boots, and threw his black pajama uniform away. He walked over and gave the jump boots to the ARVIN Soldier. The man was all smiles as he thanked him by bowing so low his chin almost touched the ground. Then he tied the boots together and hung them around his neck. The ARVIN Soldier had not worn shoes for over two years, and he did not want to get them dusty.

Breakfast tested great. After three days of running 10 miles, each day and no sleep Gilbert thought it was one of the best meals that he had ever eaten in an Army mess hall. Breakfast consisted of scrambled powered eggs, real bacon, fried potatoes, powered milk, coffee, and instead of toast, they had made pancakes with real artificial maple syrup. At Gilbert insistence, they boiled a handful of rice and flavored it with bacon not very well done. It was for the ARVIN Soldier who had been on a starvation diet for the past two years. He was allowed to drink as much of the powered milk that he wanted. He tried one glass and that was it. They already had two experience of the man's uncontrolled crapping in his pants, and Gilbert did not want another experience like that when they flew him over to the USAF Airbase to meet his wife and family.

The Montagnards sat together, the ARVIN was at a table by himself, and Gilbert took a table where he could keep an eye on all of them without appearing too. A few minutes after they started breakfast the Senior Advisor and his Adjunct, an Army Captain, walked into the mess hall and greeted him.

The Senior Advisor drank a cup of coffee while he asked, "What happened to you during the B52 bombing mission?" Gilbert answered, "no one can hear us, and the Montagnards don't understand English. The ARVIN Soldier is a little confused, and his mind is mixed up. He has been in a Tiger Cage for the past two years, and I don't think he even knows what year this is. The USAF B52 bombing mission really saved the day. I cannot say more as I am working directly under orders from Mr. Colby. We need a helicopter ride to the USAF airbase operations building, and I need a booking on an Air America flight to Saigon today." "Bill Colby asked me to tell you that he is expecting you tomorrow at 1000 hours. He has phoned me everyday since you first arrived in Plei Ku. I will phone and let him know that you will be leaving at 1500 hours and will be in his office at the time he requested. Is there anything else I can do?" The Senior Advisor asked.

"Here is a list of the PRU Tiger Scouts that were killed on my last mission. All of them were KIA (killed in action), and I would like the Family Compensation to be paid before I leave for Saigon. I'll also need a 100 RVN Gallantry Cross w/Palm, and 100 South Vietnamese Flags," Gilbert said. The Senior Advisor turned to his adjutant and said, "Get everything ready and bring it to Mr. Kelly at the Air America Passenger Counter before he leaves." "We will need the beneficiary to sign before the money is dispersed," the Adjutant answered. "Sir, please phone Mr. Colby and inform him that I'll miss his meeting tomorrow but I'll report to him when the paper work is completed," Gilbert answered. The Senior Advisor laughed, "I'll bring the funds and the other things you ordered to the airport by 1500 hours. I assume that you have someway of making sure that the compensation is delivered to the right people." "Those six Montagnard soldiers over there will take care of that little problem." Gilbert answered.

As they left the mess hall walking toward the Army Operation's building to board the helicopter taking them to the USAF Base the Montagnard Interpreter asked, "Would you have really stayed if they had not paid the money?" "Yes, I would have. You may still have to take me with you if they don't show up with the money," Gilbert answered.

When the helicopter landed in front of the USAF Operations building, there was a group of Vietnamese waiting. One was the ARVIN Army Commander and another was the Company Commander of the ARVIN Solider. Then there was his wife and she who carrying a three-month-old child in her arms while two other small children held onto her skirt. Gilbert came to attention with his PRU Tiger Scouts and all saluted. The ARVIN Commander pined a medal on the returning soldier and then they left. The soldier bowed to his wife and she returned it. Then they both bowed together. He met each child who bowed to him, and the baby held onto his finger. Gilbert watched and from the actions of the ARVIN Soldier, he thought to himself that the man really likes the baby. With out thinking but by pure instinct Gilbert called his Tiger Scouts to attention, reached into his backpack, took out three $100 Dong bills, and stepped forward giving one to each child. The mother took them from the children and bowed to Gilbert while holding the baby with the other two small children peeking out from behind their mother skirt. Then the Soldier bowed to Gilbert, and smiling started walking down the street toward the gate of the USAF Airbase. His family followed close behind him.

As the soldier walked off Gilbert knew that his work in Vietnam was doing some good. When he had first arrived in Vietnam, the men walked behind the women who followed the children as they led the way. If there were booby traps or bombs in the path or road, the children would cause them to explode or if they missed, triggering them the wife walking by would cause them to blow up. Now the roads were safe, and the man walked in front while the family followed behind. He looked grand and walked proud while stepping out in front leading his family with the boots hanging around his neck. The ARVIN Soldier was finely going home.

Gilbert looked at his watch and saw that he still had two hours before his Air America flight left for Saigon. The Senior Advisor had not arrived yet but Gilbert felt confident that the money and medals would be delivered on time. He called his Montagnards together and gave each one a hundred dollars in American script. Gilbert asked the Army helicopter crew chief to take the Montagnards to the AFX so they could buy presents to take home. Gilbert had arranged for the Army helicopter to fly them to the Kon Tum Firebase just before he left for Saigon.

At the end of every four months in Vietnam, the CIA would allow Gilbert to fly home for three weeks of R&R (leave) as Mr. Joe Kelly citizen. It was enough to keep his feet on the ground and to remember that a mortar round or rocket coming through the roof was not a fact of life.

He could not tell his friends what he did for a living. It was not a secret but the people were beginning to believe the Communist's organization "Vietnam Veterans Against the War" (VVAW) that America could not win. The news media and TV were all about Jane Fonda interviewing American prisoners in the Hanoi Hilton and leaders of the antiwar movement. Some Americans even started to follow a young Navy LTjg as he carried an Anti-Christ Peace Sign about as spokesman and member of the steering committee of the Communist VVAW with Al Hubbard. If the news media learned that Gilbert had been able to turn the Viet Cong against the Communist in his Province, it would have hurt the antiwar movement, and Joe's life would have been in danger even in the United States. Without the LTjg's support, the Vietnamese Communist would not have had a chance of getting the American people to abandon South Vietnam. There is a photo displayed in the War Remnants Museum in Ho Chi Minh City (Saigon) of the ex Navy LTjg and the former General Secretary of the Communist Party Do Muoi honoring heroes who helped the Vietnamese Communist win the war against the United States.

During this same period Secretary McNamara was using a body count of Vietnamese killed to tell the people how successful the war was going. All it did was help the Communist's plan to get the American people to give up and abandon all Christians in Vietnam. Joe had begun to believe that there was a Dark Force that used the Communists in Vietnam to attack Christians, and he knew in his heart that there must truly be a Bright Force protecting America.

In February 1971 Mr. William Colby, CIA Directorate of Plans (MACORDS), met with Joe Kelly, citizen, who was leaving the next day for home at the end of a two and a half year contract as a CIA Contract Agent and US Army South Vietnamese National Police Advisor. He had sent Joe a note asking to meet with him at the MACORDS headquarters on the Tan Son Nhut airport in Saigon. At the meeting, Joe briefed him on what had really happened at Plei Ku and Colby told him that many of the programs he was in charge of were being transferred to the US Army but that his work of helping the Vietnamese people was too valuable to be dropped.

Colby said, "Gilbert is being retired. I would like you to come back to Vietnam, continue working for the US Army, and the Vietnamese people." "It feels good helping, but I must confess that the incoming mortar shells at night are a pain in the ass, and not getting any sleep doesn't help. If the CIA is turning my work over to the US Army I will not be needed," Joe answered and then he added, "Besides the Viet Cong still has a price on my head."

The next day as he waited for every one to board the Freedom Plane (airliner) for the flight home he thought about some of the men who had helped him in the fight with the Dark Force.

One was Captain Geoffrey Barker who saw three men in dark suits wearing sunglasses that came walking down the airliner's isle; stopping at his seat. They took him and his luggage off the Freedom Bird. It was the last chance the Communists had of trying to kill him using their laws while some America's were trying to pretend that this really was not their war, and they should all go home.

Joe could not help himself, and he kept looking for the dark suited men. The airliner's cabin door was closed, and he heard the engines start. The plane started to taxi for take-off while the fasten seat belt sign came on, and no black suited men had came on the plane. The plane lined up on the runway and leaped forward. Joe could feel the nose as it pointed up to heaven, and the wheels left the runway. A loud cheer sounded from every throat; they were on their way home.

After returning home Joe Kelly, citizen, attended college, went on church missions, and eventually retired. As a hobby, he started building powered parachutes, and flying off farms near his home in Florida. It's a small world for he met another powered parachute pilot who had assisted him in Vietnam. Gentle Ben the Army Aviator who had helped move his PRU Tiger Scouts by helicopter on snatch and grab raids when he was Gilbert the CIA Interrogator. Now Gentle Ben owned a single place Buckeye 502 Powered Parachute, and joined Joe with his friends flying their powered parachutes off farmers' fields. It was exciting and pure enjoyment, just the sort of hobby that free men and women would turn too.

During February 1970, Gentle Ben had returned to Ft Rucker for treatment of injuries he received in a firefight at LZ Vivian, and upon being released from the Army hospital was sent to Washington D.C. to give a briefing to the Senate Armed Services Committee. After the meetings, he flew to San Francisco and took a bus to Oakland, California for a flight back to Saigon. As he walked into the operations center, he noticed a sign over the scheduling board for CWO Ben Games to phone a number in Washington D.C… It was his control operations center. When he called the number they asked where he was, and then told him over an unsecured line that there was no need for him to return to Vietnam as the United States was removing their advisors. Gentle Ben's mission had been canceled.

His next assignment was to the Michigan Army National Guard, and a few months later, he became a student in the Advanced Aviation Warrant Officer Course at Fort Rucker, AL as a CW-4. While tending classes Gentle Ben was also asked to write secret reports on how the North Vietnamese were going to react to proposals that the Americans were going to put on the table at the Paris Peace Talks. The Army would fly him to Washington D.C. to brief the Senate

Armed Service Committee and the Joint Chiefs of Staff. Then back to classes at the Advanced Warrant Officers College.

Joe and Ben both believe that the war against the Dark Force had started the day Jesus Christ the Son was born and would continue until Christ returned to earth. To save American and the world from sapper or terrorist attacks the Bright Force must prevail, and America must never again give up the fight for freedom or abandon the people of earth.

One day under a bright Florida sun Joe and Gentle Ben were standing together in a farmer's field after landing, and while packing their parachutes Joe turned and asked Ben, "How do you justify what we did in Vietnam?" "I believe that God does not lie. He has a plan, and we are part of it. I have been tested many times and have never been afraid to stand before my father in heaven," Ben answered and then added, "If God promised someone that he will receive seventy-two virgins in heaven if he kills a Christian then it is my job to help God to keep his promise. The sooner I can help a terrorist leave this world the happier he should be. If the man meets me in the hereafter then he will thank me or he will be so busy he will think its hell. If he gets to heaven before he kills a Christian then he will know it's really heaven, and I've been his friend."

When God created the Garden of Eden, he allowed Adam and Eve to make it their home. In the garden, there was a serpent that showed Eve a special apple and told her it would make Adam very happy. Now Vietnam was very much like the Garden of Eden, and it too had a Serpent who praised Communism while telling women it would care for their children. Like the original Adam and Eve after partaking of the forbidden fruit, they too were told that if they did not give up their religion they would have to leave Vietnam. The Dark Force controls the serpent that lies to God's children.

As the Americans fled Vietnam all the Christians and families who had fought for the Bright Force against the Communist Dark Forces were helped in fleeing to new homes in the United States. The women of the Vietnamese Families took their children and with their husbands moved to America. It was not the same as the Garden of Eden but it was a Bright Land where families could worship God and were free to choose their own path.

Senator John McCain (R-Arz) a North Vietnamese prisoner of war in the infamous Hanoi Hilton prison was finely released and returned home in March 1973. He was hospitalized for injuries received from over five years of torture by the Communist North Vietnamese. Distinguished Flying Cross for heroism awarded. Purple Heart issued.

Ben R. Games, PhD, CW-4 US Army Aviator, (Ret) returned home on the 5 February 1970 to the FT Rucker Army Hospital due to combat injuries receive in a firefight at LZ Vivian. Distinguished Flying Cross for heroism awarded. Combat missions were classified. No Purple Heart issued.

Gentle Ben's journal, records, and combat photos were transported to Fort Rucker, Al. via diplomatic courier. He was retired from the US Army 11 February 1987.

Lt/Col Geoffrey T. Baker, US Army (Ret) returned home 22 September 1970. Charged by South Vietnamese Military Tribunal for murder and torture of two prisoners. Investigation of charges proved them false and they were withdrawn. The Bronze Star was awarded for valor.

David L. Walker, Sgt. US Army Infantry, Vietnam's one year tour ended, and he was returned home March 1968. Firefight LZ Burt III Corps. The Bronze Star was awarded for valor.

The Freedom Bird (airliner) landed at Anchorage Alaska International Airport and all baggage was searched for pictures, and souvenirs of combat. David Walker saved a few pictures that were with his travel orders.

Joseph B. Kelly, CIA Contract Agent, returned home and was in a body cast from combat injuries received in an ambush near Tuy Hoa, Vietnam. Returned home on 2 December 1971. All medical reports were censored and classified. The CIA contract was for 2½ years with R&R (leave) every four months. The South Vietnamese Police Medal of Honor and the Vietnamese Pacification Bronze Star was awarded. Due to the classification of the combat as secret. No Purple Hearts were awarded.

A copy of all the documents and combat photographs while he was working as a CIA Contact Agent and as an Advisor for the US Army were stored in the United States by Joe every four months while on R&R (leave). Joe also kept a journal (diary) of his day-to-day work as Gilbert H. Morigga the CIA Interrogator.

CIA CONFESSION'S HISTORICAL DOCUMENTS

"Confession of a CIA Interrogator" is a nonfiction story of a CIA Contract employee who was assigned to the US Army as a Police Advisor during the Vietnam War for two and one half years.

There is no plot or conclusion to the story, and it was not written to change history, but only to record what Joseph B. Kelly did in it. For those students of history who disagree or just can't believe it really happened should start by reading, researching, and studying the Bible's book of Genesis, and the documents named "CIA Confession's Historical Documents".

Any reader may verify for themselves by studying the "CIA Confession's Historical Documents" what part of the story is a historical fact and what part maybe just imagination. The documents and photographs are not classified but only help support the author's theory that secrecy was used by each military service to protect their turfs from each other.

A copy of these documents and photographs are available by contacting "Little Big BOOKS ©" at:

www.GamesLittleBigBooks.com

"CIA CONFESSION'S HISTORICAL DOCUMENTS"
Little Big BOOKS ©
814 Church St, Suite 104
Ellenton, FL 34222
or the
Publisher

HISTORICAL DEDICATION PAGE

Earl R. Kelly, Staff Sgt, US Army 1945
1938 to 1941
Maryland National Guard
Co. "D" 115 Inf. 29the Division
1942 to 1941
Co. "I" 502 Praa Inf, 101 Airborne Division
Normandy, 6 June 1944
3rd Battalion 101 Airborne Division
Holland Sept 17, Bastagone Dec 17
Germany, Austria
Discharged 25 January 1945

Brother of Joseph B. Kelly, CIA Contract Agent
Vietnam 1968 to 1971

CIA CONFESSION'S HISTORICAL DOCUMENTS

Author's Statement

Prologue

Historical Documents

Photographs

www.gameslittlebigbooks.com

These documents were used by the author to help verify the story's truth & historical facts

Vietnam Pictures: Helen M. Games, MBA, Joseph B. Kelly, CIA, & Dave L Walker, Sgt.

Helen & Ben Gemes   Saigon Vietnam in route to Da Nang

STATEMENT

The story "Confession of a CIA Interrogator" is a nonfiction adventure biography of a CIA Contract Agent who was assigned to the US Army as a Police Advisor during the Vietnam War. The story covers a period of 2 1/2 years during the time William Colby headed the CIA efforts in Vietnam.

The author spent two years researching documents, photographs, and interviewing people in the story trying to verify the true facts. There were many documents located about the lost war, and after studying these along with notes taken during interviews of people who had affected the work of Joseph B. Kelly, CIA Contract Agent, it became apparent that truth is often stranger than fiction.

The author used the documents that were found to locate individuals who would talk about CIA Contract Agents and their work. Many are still afraid to have their names known. They are veterans of the Vietnam War and victims of cyber-attacks the enemy is waging against all Americans.

Ben R. Games, PhD

HISTORICAL PROLOGUE

South Vietnam has a temperature averaging 80 degrees throughout the year with an annual rainfall of 79 inches. Fuuj in's summer monsoon winds bring the rains down from the north cleaning and sweetening the air. The water runoff from the Tropical Rain Forests feed the Mekong River making its many mouths of delta land the richest rice-growing area in the world.

Into this land came three people unlikely to ever meet, but destined to learn from each other. One man was a young US Navy LTjg Skipper of a Swift Boat, one was an Army Chief Warrant Officer, Ben R. Games, Chinook Aircraft Commander, and one was a CIA Contract Agent Joseph B. Kelly who became Gilbert a CIA Interrogator. Their work effectively helped the Communist Viet Cong (terrorists) to destroy themselves.

During WW II, this peaceful land was known as French Indo China, and was invaded by the Imperial Japanese Army. A few years after Japan was defeated the United Nations divided the Nation into three countries called Laos, Cambodia, and Vietnam. France was given control of Vietnam until the Communists forced them to retreat to the southern part of the country.

The South Vietnam capital of Saigon was a beautiful city, and it soon became known throughout the world as the Paris of the Orient. Only ten percent (10%) of the population was of the Christian faith with all civil service employees, ministers, and leaders being Catholic. The population of South Vietnam in 1968 during the time Joseph B. Kelly became Gilbert the CIA Interrogator was 15,715,000 people living on 65,726 square miles of land (similar in size to the state of Washington).

The land was controlled by two forces. The Dark Force known as the Viet Cong (VC) under the command of the Communists "Peoples Revolutionary Party Central Committee (PRP)" of one man and two women supported by the North Vietnamese, and the Bright Force under the command of their elected President Thieu supported by the Americans. The Dark Forces controlled the country after the sun went down. They set Booby Traps on trails, roads, and bombed government facilities. There were VC Communist tax collectors with police and courts. The nights belonged to the terrorists, and during daylight, the Bright Forces took control.

When Gilbert the Interrogator was sent to the Mekong Delta region, he was given a mission to assist the National Police in establishing an intelligence system to find and locate terrorists. To accomplish this mission all the Dark Force Tax Collectors and other officials of the Communist Viet Cong had to be identified. He desperately needed to capture and identify the enemy, not kill them. Gilbert could not interrogate or find out where the VC were going to attack if they were dead, and nothing was gained if the Viet Cong were allowed to commit suicide.

Most of the 720 young men and women who worked for Gilbert in the CIA secret Dai Phong Program were North Vietnamese. They had volunteered to join the Province Recon Units (PRU) Tiger Scouts attempting to stop the VC terrorist attacks. As members of the PRU Tiger Scouts, the men were paid a monthly salary of about two hundred dollars, and the Communist Viet Cong women were paid as informers. They received $25 for each weapon cache and $50 to identify a VC official.

Despite the overuse of classifying everything secret by the State Department and the military services there was still an abundance of unclassified documents and photographs left over. When these are studied and put together, they paint a picture of true chaos. The US Army with the intelligence provided by the CIA could have won the war without the help of the other services if let alone. In fact, the US Navy could have done the same thing, and the USAF knew that they could have. If the politicians had wanted to, they could have hired the Koreans to pacify South Vietnam and forgot about the war.

The American military Veterans of the Vietnam War did not lose the battle or give up. When they returned home, they should have been treated as the heroes that they truly were. Instead, they were treated as if they had done something wrong and unclean. Today sociologist know that is caused by guilt from leaving the Vietnamese people to die. It is part of the same guilt that Americans feel because of the African Americans who were brought to the United States as slaves yet there are no people alive today who captured or owned an African American slave. Maybe if the Vietnam Veterans would forgive President Kennedy, President Johnson, and President Nixon it would help the Americans have closure.

13. EXPERIENCE (Start with your PRESENT position and work back.)

May inquiry be made of your present employer regarding your character, qualifications, and record of employment? **YES** / **NO: X**

1.
- Dates of employment: From 5 Aug. '68 To PRESENT TIME
- Exact title of position: Intelligence Officer
- If Federal service, classification series and grade: GS-11
- Salary or earnings: Starting $12,000 per year; Present $12,699 per year
- Avg. hrs. per week: 40
- Place of employment: City: Washington; State: D.C.
- Number and kind of employees supervised: 36 Vietnam
- Kind of business or organization: U. S. Government
- Name and title of immediate supervisor: Director of Personnel, Attn: V. L. Surry
- Area Code and phone No. if known: (703) 351-3295
- Name of employer: Central Intelligence Agency, Washington, D. C. 20505
- Reason for wanting to leave: End of Contract
- Description of work: Served both in the United States & Southeast Asia as an Intelligence/Administrative Officer. In Southeast Asia and while detailed to another government agency, was the officer in charge on various programs which included liaison with and guidance to local police organizations, and supervision of various other programs and projects of the US. Government. The management of these programs involved direct supervision of a large complement of indigenous personnel and American subordinates, complete accountability and control of funds, and full responsibility for the logistical support of the program.

2.
- Dates of employment: From 1 Jan. '48 To 1 Jul. '68
- Exact title of position: Survival Personal Superintendant
- If Federal service, classification series and grade: Sgt.
- Salary or earnings: Starting $6,000 per year; Final $ per
- Avg. hrs. per week: 40
- Place of employment: City: ; State: USAF
- Number and kind of employees supervised: 25 Military
- Kind of business or organization: Government
- Name and title of immediate supervisor: Col. Nugent, Commander
- Area Code and phone No. if known: Unknown
- Name of employer: United States Air Force
- Reason for leaving: Retirement
- Description of work: Supervisor of high altitude flight suits and gear used in U-2s and B-57's. This was used with high reconnaissance air craft. Instructed in first aid and survival for the Squadron. Was the NCO Advisor to the commander. Also instructed in the proper use of Parachutes and floatation gear used in the aircraft of the Squadron. Instructed in scuba and water survival. Tested all high altitude pressure suits in special chambers of aero space and in the aircraft they were going to be used with. During part of period served as Atomic Release Officer responsible for insuring proper security procedures were met prior (Continued on #34).

3.
- Dates of employment: From Sept. '61 To Feb. '62
- Exact title of position: Security Agent
- Salary or earnings: Starting $ per; Final $ per
- Avg. hrs. per week: 8
- Place of employment: City: Kansas City; State: Missouri
- Kind of business or organization: Security
- Name of employer: Pinkerton Detective Agency, Kansas City, Missouri
- Reason for leaving: Military transfer
- Description of work: Various security duties during part-time employment for private detective agency.

IF YOU NEED ADDITIONAL EXPERIENCE BLOCKS USE STANDARD FORM 171-A OR BLANK SHEETS SEE INSTRUCTION SHEET

Viet Cong Choi Hoa Pamphlet for Americans

The Anti-Christ Peace Sign

THE ATLANTIC MONTHLY

TOUR OF DUTY

John Kerry in Vietnam

BY DOUGLAS BRINKLEY

*Atlantic
monthly
Dec 2003*

As he campaigns for the 2004 Democratic presidential nomination, Senator John F. Kerry often cites his experien ~ as a U.S. Navy patrol-boat skipper in Vietnam as a formative element of his charact ~h the historian Douglas Brinkley will publish the first full-scale, intimate account c ~writing that account Brinkley has drawn on extensive interviews with ~well in Vietnam, including all but one of the men still l ~d over to Brinkley his letters home from Vietna ~and personal reminiscences written c ~hout restriction, to be used

John Kerry enliste
In December of 196
after five months of
States and underw
rivers. In June of

Douglas Brinkley, a historian and biographer, is the director of the Eisenhower Center for American Studies and a professor of history at the University of New Orleans. His most recent book is Wheels for the World (2003), a biography of Henry Ford. Tour of Duty: John Kerry and the Vietnam War *will be published by William Morrow in January.*

sufferec
Star for gallantry
of 1970, and soon became
antiwar movement.

The following excerpts are drawn from D~
John Kerry and the Vietnam War.

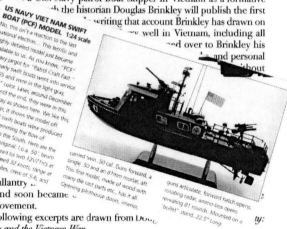

Photo of John Kerry and Vietnam's former General Secretary of the Communist Party Do Muoi. The photo is part of an exhibit honoring heroes who had helped the Vietnamese Communists win the war against the United States. The photo is displayed in the War Remnants Museum in Ho Chi Minh City.

The Enemy

The Enemy:

+++ Is communist North Vietnam and its southern arm, Viet Cong (Vietnamese communist or "Viet Red").

+++ Controls, directs, supplies entire effort to conquer Republic of Vietnam through COSVN (Central Office for South Vietnam), which heads military, political efforts of North Vietnam in South Vietnam, closely resembles government structure, reaches into every district.

+++ Trains military, political cadres, terrorists, spies, saboteurs, providing most of VC leadership. These have infiltrated in increasing numbers since 1956.

+++ Uses systematic terrorism, assassination to wreck economy, destroy fabric of government of 2,500 villages. In past five years communists have assassinated, beheaded or kidnaped 2,000 village chiefs, have driven away able-bodied men, have deluded or terrorized many citizens into cooperating.

+++ Still failing to win, in 1965 Hanoi began sending regular units of North Vietnamese Army to south. Backbone of communist military in South Vietnam is VC "main force" and NVA units, of about 112,000. VC guerrillas number 119,000, political cadres 39,000, combat support 18,000. (Estimates are as of summer 1966.)

+++ Estimated average of at least 5,000 men infiltrated South Vietnam monthly during first seven months of 1966, plus supplies, by land and sea from north. Arms are from China, Russia, Red satellites.

+++ In first half of 1966, however, 10,000 VC and NVA soldiers and sympathizers turned themselves in to South Vietnam government under Chieu Hoi (open arms) program.

Navy—Pearl Harbor

The Stakes In Vietnam

Free World Goals:

+++ Freedom and independence for 15.5 million South Vietnamese.

+++ Protect all Southeast Asia by stopping communist aggression in South Vietnam. (Red's announced next target: Thailand.)

+++ Peace for Free World by defeat in Vietnam of communist plans for "wars of national liberation" (euphemism for externally inspired, directed and conducted aggression against a government). U. S. Commitment:

+++ Three U. S. Presidents have pledged assistance to South Vietnam.

+++ Southeast Asia Treaty Organization (SEATO), of which U. S. is member, pledged to protect area against communist aggression.

+++ Annually since 1956 U. S. Congress has voted economic, military assistance programs for Vietnam.

Wider Significance:

+++ Security of U. S. and rest of Free World is involved, not only that of Southeast Asia, because communists are testing techniques, tactics in South Vietnam.

+++ To leave South Vietnam to its own defenses would shake confidence of millions who respect and value U. S. commitments, promises.

+++ President Johnson: "We will stand in Vietnam."

John Kerry and Al Hubbard on Meet the Press in April 18, 1971. Kerry was asked if he himself had committed any war crimes or atrocities in Vietnam. Kerry answered that he had (partial transcript on page 153).

CIA Contract Agent
&
PRU Tiger Scouts

Hanoi Jane's Apology

by: Bruce Herschensohn

The Washington Times—Americans are in a hurry to be done with the past and go on to tomorrow. As a clear example of that, pick up the July-August edition of "O: The Oprah Magazine."

That magazine includes an interview with **Jane Fonda** with an introduction by **Oprah Winfrey**.

Ms. Winfrey writes that Jane Fonda is "the same Jane who protested the Vietnam War and made some Americans so angry that they labeled her a Communist and slapped her with the nickname of Hanoi Jane." Either Ms. Winfrey doesn't remember or didn't know that the reason "some Americans" thought she was a communist came from direct statements of Ms. Fonda.

On Nov. 21, 1970 she told a University of Michigan audience of some 2,000 students.

"If you understood what communism was, you would hope, you would pray on your knees that we would some day become communist."

At Duke University in North Carolina she repeated what she had said in Michigan, adding "I, a socialist, think that we should strive toward a socialist society, all the way to communism."

She didn't merely protest the Vietnam War, as Oprah Winfrey wrote. Jane Fonda took the side of the North Vietnamese. In that recently published interview Jane Fonda states,

"I will go to my grave regretting the photograph of me in an antiaircraft carrier, which looks like I was trying to shoot at American planes. That had nothing to do with the context that photograph was taken in. But it hurt so many soldiers. It galvanized such hostility. It was the most horrible thing I could possibly have done. It was just so thoughtless. I wasn't thinking. I was just so bowled over by the whole experience that I didn't realize what it would look like."

It appears to me as though Jane Fonda is sorry about the photo, but she is not apologizing for her actions that led to the photo since "the context" of which she speaks is by far worse than the photograph. That photo was taken when she went to North Vietnam in July of 1972 where she not only posed for a photo, but also recorded propaganda broadcasts for the North Vietnamese.

Among her statements are these precise quotes: "I'm very honored to be a guest in your country, and I loudly condemn the crimes that have been committed by the U.S. Government in the name of the American people against your country. A growing number of people in the United States not only demand an end to the war, an end to the bombing, a withdrawal of all U.S. troops, and an end to the support of the **Thieu** clique, but we identify with the struggle of your people. We have understood that we have a common enemy: U.S. imperialism."

And: "I want to publicly accuse **Nixon** of being a new-type **Hitler** whose crimes are being unveiled. I want to publicly charge that while waging the war of aggression in Vietnam he has betrayed everything the American people have at heart. The tragedy is for the United States and not for the Vietnamese people, because the Vietnamese people will soon regain their independence and freedom...."

And: "To the U.S. servicemen who are stationed on the aircraft carriers in the Gulf of Tonkin, those of you who load the bombs on the planes should know that those weapons are illegal. And the use of those bombs or condoning the use of those bombs, makes one a war criminal."

And: "I'm not a pacifist. I understand why the Vietnamese are fighting...against a white man's racist aggression. We know what U.S. imperialism has done to our country so we know what lies in store for any third world country that could have the misfortune of falling into the hands of a country such as the United States and becoming a colony...You know that when Nixon says the war is winding down, that he's lying."

Within six months our military involvement was over.

I was working for President Nixon at the White House when our men returned from being prisoners of war and I talked with many of them. For refusing to meet with her, a naval commander was beaten daily while in a three-foot by five-foot windowless cell, held there for four months. A lieutenant commander was hung by his broken arm attached to a rope, then dropped by the end of the rope time after time as the table he stood on was kicked out from under him.

A captain was hung under his elbows from rounded hooks on his cell wall and beaten into unconsciousness with bamboo sticks.

Here are a few of the direct quotes that I saved from those days:

Lt. Comdr. John McCain said, "These people, **Ramsey Clark**, **Tom Hayden**, and **Jane Fonda**, were on the side of the North Vietnamese. I think she only saw eight selected prisoners. I was beaten unmercifully for refusing to meet with the visitors."

Maj. Harold Kushner said, "I think the purposes of Fonda and Clark were to hurt the United States, to radicalize our young people, and to undermine our authority."

Col. Alan Brunstrom said, "We felt that any Westerners who showed up in Hanoi were on the other side. They gave aid and comfort to the enemy, and as far as I'm concerned, they were traitors."

After the U.S. prisoners of war returned and had landed at Clark Field in the Philippines in 1973, Jane Fonda publicly said that they were "hypocrites and liars and history will judge them severely."

Jane Fonda has now apologized for a photograph, but she speaks about some unexplained context. The context is the crime. The photograph is merely the visual evidence of the crime.

This article was mailed from The Washington Times.

The Body Count

Ohio State University Woody Hays, Football Coach Gilbert, CIA Interrogator Marty Karow, Baseball Coach

Visit to Firebase in VC Country Cao Lanh, 4th Corp Area Cambodian Border 1969

TRANSLATION

Vietnamese Citation for award of Bronze Star to English

for Joseph B. Kelly (1970)

REPUBLIC OF VIETNAM
KIEN HOA PROVINCE
PHUNG HOANG COMMITTEE
PERMANENT CENTER

TO: LT COLONEL PROVINCE CHIEF
CONCURRENTLY KIEN HOA SECTOR COMMANDER

SUBJECT: Propose a official award for an American Advisor and his assistant

Dear Sir,

I respectfully report to you that:

1- **Mr Joseph B. Kelly, PIC ADVISOR:**
Since the arriving day to this province up to date, he has shown proof of good will and outstanding advisor. He launched the DIA PHONG campaign in this province and gave it a very effective support, resulted over 300 overt and covert VCI were neutralized.

Beside that, MR KELLY also was a brave advisor. He participated in many Police and PHUNG HOANG OPERATIONS, particularly on 26 May 70 he conducted a National Police Field Force Platoon to operate at PHU DANG hamlet, THANH THOI village, Mo Cay district, resulting in 5 VC killed, one M3A1 sub machine gun, two anti tank mines and many ammunitions of all types were seized.

With the above brilliant feats, I would like to propose you to award him a Cross of Gallantry with Bronze Star Medal (commended before Regiment) on the occasion of his transfer to another province in July 1970.

2- **Mr Pham Van Dien, Assistant PIC Advisor:**
Side by side with Mr Kelly in all missions, Mr Pham Van Dien has a special talent to proselyte ex VC cadre to work for the Dai Phong campaign. He also provides to PHUNG HOANG COMMITTEE many good information to eliminate VCI.

For this reason, I would like to propose you to award him a Certificate.

Kien Hoa, 26 June 1970
Major Ho Van Man
Phung Hoang Center Chief
S/S

Korean convoy escorts Quat 50's

Recon by fire

On the Beach for a party. Gentle Ben, Helen, Korean Embassy Military Attaché, Korean Intelligent Officer.

Mariene air assault (UH!).

REPUBLIC OF VIETNAM
MINISTRY OF INTERIOR

Kien Hoa 25 April 1970
No. 002636/CSKH/CSDB/KH-M.

DIRECTORATE OF N.P. IN IV CORPS
NATIONAL POLICE SERVICE
KIEN HOA PROVINCE

TO : OSA Director in IV Corps, Phong-Dinh

FROM : National Police Service Chief, Kien Hoa

SUBJECT : Request for extension for Mr. J. B. KELLY to serve one more year in Kien Hoa Province.

My Dear Sir,

Mr. Joseph B. Kelly, PIC Advisor, arrived Kien Hoa Province since June 1969. Beside the routine work as advising my PIC chief to operate properly, M. Kelly has proselyted and persuaded the captured VC cadres to cooperate with the GVN to eliminate VCIs in the province town. The above VC cadres is formed in small teams and combined with Special Policemen to conduct DAI PHONG operations to capture VC city penetrators. These operations have produced good results because from September 1969 up to date, over 150 VCIs were captured. They are:

- Sappers
- Military Intelligence Agents
- Military Proselyting Cadres
- Security Cadres
- Propaganda & Training Cadres
- Commo-Liaison Agents
- Armed Security Cadres

Beside the cooperation with the Police to eliminate VCIs, Mr. Kelly has also organized an English course for policemen to improve their knowledge and to tighten the Vietnamese-American friendship. He works day and night and devotes all his time to help not only the PIC but also the National Police Service.

With the good results produced by DAI PHONG operations as stated above, we plan to expand the DAI PHONG program to 9 districts in this province to eliminate all VCIs from Province town to districts.

Now, it is known that Mr. Kelly will finish his service in Viet Nam at the end of June 1970. Therefore I would like to request you to interfere with American Embassy to grant him permission to stay one more year in Kien Hoa province.

So far, we have had many PIC Advisors, but none of them did a very good job as Mr. Kelly. Evidences are he was issued a Certificate by the Province Chief and proposed a Police Medal of Honor for his outstanding performance.

If Mr. Kelly is extended his service in Kien Hoa Province for one more year, we believe that the DAI PHONG program will reap substantial results by his support and good cooperation.

Respectfully Yours,

By order of N.P. Service Chief

Deputy

HUYNH CAO PHAM

s/s.

Copies to:

- Colonel Director General of NP, Saigon
  (Special Police Division)
- Lt. Colonel Director of NP in IV Corps, Phong Dinh (Special Police Office)
  "for info"

PRU TIGER SCOUTS
Cao Lanh, Vietnam

Province Recon Unit Training (PRU Tiger Scouts)

Gilbert & Cambodian Snatch & Grab Team

PRU BOATS

14′ Starcraft Boat made in Goshen, IN. Gilbert making plans for beach party at Tuy Hoa AFB.

VC Sappers
Blew up bridge

PRU "MONITER"
Mekong River

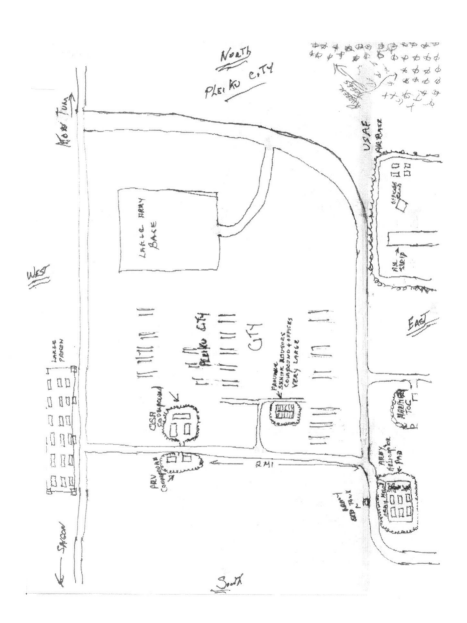

Meeting of ARVIN Province Chief & Officers.
Gilbert is on the far right.

USAF Lt/Col Peters B-57F pilot
Joining the Montagnard Tribe Six by drinking 6 sticks of wine

PHOTO BY 1LT RALPH F. CAMPBELL

(Top) Soldiers scurry from choppers toward cover. (Bottom) Infantrymen prepare to jump out of the chopper and assault a wood line.

PHOTO BY SP4 JOSEPH CAREY

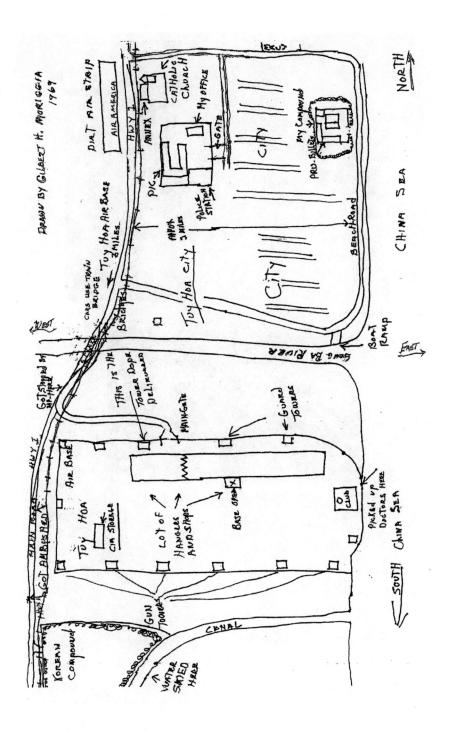

U.S. ARMY PHOTO
A Ch-47 (Chinook) carries a sling load of supplies to men in the field.

RECON BY FIRE

REPUBLIC OF VIETNAM
The Interior Department
--/--
DECREE NO. 191/BNV/VP/1/ND

On 24 March 1970 a Decoration ceremony was held to pin the Police Meritous and Honorable Medals to 86 Policemen, personnel and the US Coordinators.

The Secretary of State for Interior Department

- Considering the Constitution of the Republic of Viet-Nam dated on 1 April 1967;

- [...] ted on 1/9/1969, to stipula[...]
- [...] d 22/11/1967, to stipula[...] te for the Interior [...]
- [...] on 13/7/1968; to stipula[...] erior Department;
- [...] dated on 6/9/1965, to [...] ble Medals;
- [...] ND dated on 21/1/1966, [...] ocedures of rewarding the Police Meritous a[...]
- [...] ree No.241/ND/HP/PC dated on 1/2/1967, to e[...] to the General Commissioner for Security a[...] tor of National Police to examine and approv[...] olice Meritous and Honorable medals;
- [...] ree No.299/ND/Th.T/PC dated on 27/12 1967, to r[...] .241/ND/HP/PC dated on 1/2/1967 in endorsi[...] the Secretary of State for Interior Department [...] ctor of National Police in making the decora[...] ous and Honorable medals;
- [...] ree No.782/ND/Th.T/VP dated on 27/7/1968 [...] No.299/ND/Th.T/PC dated on 27/12/1967 in en[...] y to the Minister for Interior Department [...] ice Meritous and Honorable medals;
- Considering the Decree of the General Director of National Police;

B. POLICE HONORABLE MEDAL (The third Class) be pinned to:

76.- Joseph B. Kelly, US Coordinator for the NP/KH.

Ông JOSEPH KELLY Phối trí viên cạnh TRUNG TÂM THẨM VẤN TY CẢNH SÁT QUỐC GIA KIẾN HÒA ĐƯỢC ÂN THƯỞNG HUY CHƯƠNG

Hồi 08 giờ sáng ngày 27.4.70, Ty Cảnh Sát Quốc Gia Kiến Hòa đã tổ chức lễ gắn huy chương Cảnh Sát Danh Dự đội tinh cho 1 cố vấn Hoa Kỳ và tiễn đưa 75 nhân viên Cảnh Sát tân tuyển lên đường thụ huấn tại Trung Tâm Huấn Luyện Rạch Giữa. Ông Trưởng Ty Cảnh Sát Quốc gia Kiến Hòa đã đại diện Đại Tá Tổng Giám Đốc Cảnh Sát Quốc Gia đã gắn huy chương Cảnh Sát danh dự Bội Tinh Đệ III Đẳng cho Ông JOSEPH KELLY, Phối Trí Viên Cạnh Trung Tâm Thẩm Vấn Ty Cảnh Sát Quốc Gia Kiến Hòa.

Nhân dịp này, Ông Trưởng Ty Cảnh Sát Quốc Gia Kiến Hòa cũng đã chứng dự lễ tiễn đưa 75 nhân viên Cảnh Sát tân tuyển lên đường thụ huấn tại Trung Tâm Huấn Luyện Rạch Giữa.

Số nhân viên Cảnh sát Tân tuyển trên đây thành phần gồm quân nhân ĐPQ/NQ xi biệt phái phục vụ ngành Cảnh Sát Quốc Gia tại các Phân chi trong toàn Tỉnh sau thời gian thụ huấn 1 tháng rưỡi tại Rạch Giữa.

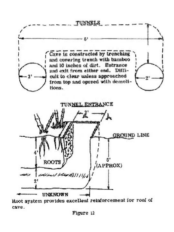

Root system provides excellent reinforcement for roof of cave.

Figure 12

DEPARTMENT OF THE ARMY
OFFICE OF THE DEPUTY CHIEF OF STAFF, G-1
2461 EISENHOWER AVENUE
ALEXANDRIA, VA 22332-0300

JUN 0 1 2006

REPLY TO
ATTENTION OF

Policy and Program Development Division

Honorable Ginny Brown-Waite
United States Representative
20 North Main Street, Suite 200
Brooksville, FL 34601

Dear Congresswoman Brown-Waite:

This is in response to your recent inquiry on behalf of Mr. Joseph B. Kelly, regarding his eligibility for the Purple Heart.

The criteria for the Purple Heart during the time Mr. Kelly was employed in Vietnam required injuries to be directly related to enemy action and were also dependent upon whether the recipient was in direct or indirect combat operations. These determinations were made by commanders and managers at the location of the combat operation. Based on review of the information provided, Mr. Kelly's injury did not meet the criteria for the Purple Heart. The Defense of Freedom Medal was not established at the time of Mr. Kelly's service in Vietnam and may not be presented for an injury prior to its establishment.

Mr. Kelly was an Army civilian employee at the time of his injury and received the Civilian Service in Vietnam Medal for his contributions to the Department of the Army. The support Mr. Kelly provided as a civilian in Vietnam was commendable. Please convey our thanks to him for his service.

Sincerely,

Jeannie A. Davis
Chief, Policy and Program Development
Division

2 February 1984

TO WHOM THIS MAY CONCERN

    In 1970 I was assigned as a reconnaissance advisor in the Republic of Vietnam. On the evening of 22 September, the residence and compound where I was assigned in the city of Ben Tre, Kien Hoa Province, came under a deadly mortar attack. One round crashed through the roof of the house into the bedroom of US Civilian employee, Joseph B. Kelly. Also present that evening, in addition to the five Americans residing in this building, was Major General (USA Ret.) Charles Timmes.
    Between securing General Timmes beneath a stairwell in a protected portion of the building, and rallying the indigenous defense force, I distinctly remember seeing Mr. Kelly. Although obviously disoriented from the blast of the mortar round that exploded in the rafters above his sleeping area, he was gallantly taking charge and accounting for the Vietnamese civilian nationals within the compound. Mr. Kelly volunteered to accompany the reaction force departing to attempt to locate the enemy position, however as appeared temporarily deafened from the tremendous noise at such close quarters, he was placed in charge of compound security and defense.
    It was several days before Mr. Kelly could participate in normal conversations. His associates were required to speak louder than normal even when in close proximity; and he in turn spoke loudly, as people with hearing disabilities tend to do. For the next two months (prior to my reassignment to a different Province, I remained with Mr. Kelly and the other residents of the Embassy House in Ben Tre. During that entire period Mr. Kelly complained he could still hear ringing noises in his ears.
    In my professional opinion, Mr. Kelly is extremely fortunate to have survived the blast of the mortar round that landed in his vicinity, and should have received a civilian award. Although clearly suffering he demonstrated selfless determination and dedication to duty.

                                          Geoffrey T. Barker
                                          Major, Infantry
                                          G3 Plans
                                          1st Special Operations Command
                                          Fort Bragg, North Carolina.

employment
4. Reporting office __EMBASSY HOUSE__ (Arsenal, navy yard, etc.) __BEN TAR CITY__ (City) __KIEN HOA__ __VIETNAM__ (State)
(Location of reporting office or division headquarters)
5. Name of superintendent or foreman in charge when injury occurred __Mr John Thomas__

6. Name of injured employee __Joseph B. Kelly__ 7. Age __39__ 8. Sex __M__ 9. Citizenship __AMERICAN__
(Give first name in full)
10. Home address __462 Wyn Ann Ave__, __Aberdeen__, __Maryland__
(Street and number) (City or town) (State)
11. Occupation and division __Ordnance Officer__ 12. Was employee doing his regular
(Give both, as laborer, hull division; helper, machine shop, etc.)
work? _____ If not, what work? __Asleep in bed in Company House__

The injured employee
13. Total length of service with the Government as a civilian? __2 yrs__
14. How long at present work in this establishment? __1 years__
15. Dates of other injuries __None__
16. Rate of pay on date of injury, $ __?__ per _____ { and subsistence valued at $_____ per _____
                                                        { and quarters valued at $_____ per _____
17. Employee begins work at __0900__ A.m. 18. Regular day's work ends __1800__ A.m.
(Hour, a.m. or p.m.) (Hour, a.m. or p.m.)
19. Hours worked per day __8__ 20. Days paid per week __5__

21. Place where injury occurred __Embassy House Ben Tre City Kien Hoa Province Vietnam__
(Give exact location, as name or number of building and division, etc.)
22. Date of injury __8 June 1970__, 19 __70__; day of week __Monday__; hour of day __01:00__ __a.m.__
23. Date employee stopped work __N/A__, 19 ___; day of week ___; hour of day ___ __.m.__
24. Date employee's pay stopped __N/A__, 19 ___; day of week ___; hour of day ___ __.m.__
25. Has employee returned to work? __NEVER STOPPED WORK__
(Give date and hour)
26. Will employee receive pay for any portion of above absence on account of:
    (a) Annual leave __N/A__
    (b) Sick leave __N/A__ (Give exact dates)
    (c) Any other reason __N/A__ (Give exact dates)
27. Describe in full how injury occurred __WAS ASLEEP IN QUARTERS WHEN 3 ROUNDS OF 82 MORTAR HIT OUR HOUSE. 1 round HIT THE ROOM I WAS sleeping in. I was at the DOOR Trying to get out when the mortar hit my room__
28. State part of body injured and nature and extent of injury __CAN'T HEAR properly out OF MY LEFT EAR. BLAST WAS 15 FEET AWAY__

The injury
29. Did injury cause loss of any member or part of member? __NO__ If so, describe exactly __N/A__
30. Was employee injured while in performance of duty? __YES__ If not, or in doubt, give detailed statement _____
31. Was injury caused by:
    (a) Willful misconduct of the employee? __NO__ (b) Intention of employee to bring about injury or death of himself or another? __NO__ (c) Employee's intoxication? __NO__
    (If any answers to these questions are made in the affirmative, the reporting officer should attach an additional statement giving the reason for his conclusion)
32. Was written notice of injury given within 48 hours? __NO__ If not, did immediate superior have actual knowledge of injury? __Yes And it was Reported by Radio to Saigon__
(Answer to question 5, Form C. A. 1, must be complete if notice was not given within 48 hours)
33. Names and addresses of witnesses to injury __F(b)(3)__
__P(j)(1)__
__(U)__

34. Was injury caused by a third party other than a Government employee or agency? __YES__ If so, has employee been instructed in procedure under the Bureau's regulations? __VIET CONG (VC)__
(A detailed statement should be forwarded with this report)

Medical attendance
35. Name and address of physician who first attended case __Dr ▓▓▓▓▓▓ MD__
36. How soon after injury? __10 Days__
37. To what hospital sent? __Embassy Dispensary__ Location __▓▓▓▓▓▓__
38. Name and address of physician now attending case __Dr ▓▓▓▓▓▓ Dispensary__

U.S. DEPARTMENT OF LABOR
WAGE AND LABOR STANDARDS ADMINISTRATION
Bureau of Employees' Compensation

EMPLOYEE'S NOTICE OF INJURY OR OCCUPATIONAL DISEASE
(Under the Federal Employees' Compensation Act)

3 copies

INSTRUCTIONS
This form should be completed by the injured employee or someone on his behalf whenever an injury is sustained in the performance of duty and given to his immediate superior within 48 hours. It should be placed in the employee's official personnel file unless the injury causes disability for work beyond the day when it occurred; is likely to result in prolonged treatment or permanent disability; or in a charge for medical or related expenses when it should be forwarded to this Bureau with Form CA-2, Official Superior's Report of Injury. This form is also completed whenever an employee believes he suffers from a disease related to his employment. (See Sections 1.2, 1.3, 2.2 and 2.3 of the Bureau's Regulations.)
The immediate superior should also complete the reverse side of this form.

1. NAME OF INJURED EMPLOYEE (Last, first, middle): Lilly Joseph Bernard
2. DATE OF THIS NOTICE (Mo., day, yr.): 22 Sept 70
3. PLACE OF EMPLOYMENT (Name and location of office or establishment): Embassy House Tuy Hoa City Phu Yen Province
4. DATE OF INJURY (Mo., day, yr.): 22 Sept 70
5. OCCUPATION: Special Police Advisor
6. HOUR OF INJURY: 1100
7. PLACE OR LOCATION WHERE INJURY OCCURRED: Vietnam Tuy Hoa City Phu Yen Province
8. CAUSE OF INJURY (Describe how and why injury occurred):
Received small arms fire over the pickup truck I was driving between Phu Hiep Army Post and Tuy Hoa Air Force Base. I speed up to avoid being hit and a small boy drove his cow across the road. I took the pickup into the ditch to avoid the cow. When I drove the truck into the ditch it turned over.
9. NATURE OF INJURY (Name part of body affected—fractured left leg, bruised right thumb, etc.):
Lower Back Pain - Fracture of the 5th Lumbar Spine vertebrae
10. NAMES OF WITNESSES TO INJURY:
Mr Cam Master Phu Yen Province Tuy Hoa City Vietnam
11. IF THIS NOTICE WAS NOT GIVEN WITHIN 48 HOURS AFTER THE INJURY, EXPLAIN REASON FOR DELAY. IF EARLIER NOTICE WAS GIVEN, VERBAL OR WRITTEN, STATE WHEN AND TO WHOM.

I certify that the injury described above was sustained in the performance of my duties as an employee of the U.S. Government and that it was not caused by my willful misconduct, intention to bring about the injury or death of myself, or another, nor by my intoxication. I hereby make claim for compensation and medical treatment to which I may be entitled by reason of this injury.

12. SIGNATURE: Joseph B Lilly
13. HOME ADDRESS OF INJURED EMPLOYEE: 403 Wyn Mar Ave Aberdeen Maryland

The Purple Heart
Then and Now

Attached to the piece of dark blue cloth is a purple heart of silk, bound with braid and edged with lace. The cloth is believed to be part of the uniform tunic of a soldier of the Continental Army.

There is no name, rank or regimental insignia on the piece of cloth. The Purple Heart is displayed in Washington, D.C., at the Society of the Cincinnati's Anderson House Museum and another at the New Windsor Cantonment site at New Windsor, NY. The Purple Heart itself is what signified a hero in the Revolutionary War.

The Purple Heart was awarded to three soldiers - Sgts. Elijah Churchill, William Brown, and Daniel Bissell, Jr. On May 3, 1783, Churchill and Brown received the Purple Heart, then called the Badge of Military Merit, from Gen. George Washington, its designer and creator. Bissell received his on June 10, 1783. These three were the only recipients of the award during the Revolutionary War.

On August 7, 1782, at his Newburg, NY headquarters, Washington devised two badges of distinction to be worn by enlisted men and noncommissioned officers. The first was a chevron to be worn on the left sleeve of the coat. It signified loyal military service. Three years of service with "bravery, fidelity, and good conduct" were the criteria for earning this badge; two chevrons meant six years of service.

The second, named the Badge of Military Merit, was the "figure of a heart in purple cloth or silk-edged with narrow lace or binding". This badge was for "any singularly meritorious action" and permitted the wearer to pass guards and sentinels without challenge. The honoree's name and regiment were inscribed in a Book of Merit.

After the Revolutionary War, no more American soldiers received the Badge of Military Merit. It was not until October 10, 1927, that Army Chief of Staff, General Charles P. Summerall, directed a draft bill be sent to Congress "to revive the Badge of Military Merit."

The Army withdrew the bill on January 3, 1928, but the Office of the Adjutant General filed all correspondence for possible future use.

Although a number of private efforts were made to have the medal reinstituted, it wasn't until January 7,m 1931, that Summerall's successor, General Douglas MacArthur, confidentially reopened the case. His object was to have a new medal issued on the bicentennial of George Washington's birth.

Miss Elizabeth Will, in the Office of the Quartermaster General, created the design from guicelines provided her. The only difference in her design is that a sprig appeared where the profile of Washington is on the present Purple Heart.

John R. Sinnock of the Philadelphia Mint made the plaster model in May 1931. The War Department announced the new award on February 22, 1932.

After the award was reinstated, recipients of a Meritorious Service Citation Certificate during World War I, along with other eligible soldiers, could exchange their award for the Purple Heart.

At the same time, revisions to Army regulations defined the conditions of the award: *"A wound which necessitates treatment by a medical officer and which is received in action with an enemy, may in the judgment of the commander authorized to make the award, be construed as resulting from a singularly meritorious act of essential service."*

At that time the Navy Department did not authorize the issue of the Purple Heart, but Franklin D. Roosevelt amended that by Executive Order on December 3, 1942, with the Coast Guard beginning December 6, 1941.

President Harry S. Truman retroactively extended eligibility to the Navy, Marine Corps, and Coast Guard to April 3, 1917, to cover World War I.

President John F. Kennedy extended eligibility on April 25, 1962, to "any civilian national of the United States who, while serving under competent authority in any capacity with an armed force ..., has been, or may hereafter be, wounded."

President Ronald Reagan, on February 23, 1984, amended President Kennedy's order, to include those wounded or killed as a result of "an international terrorist attack."

Army regulations, amended June 20, 1969, state that any "member of the Army who was awarded the Purple Heart for meritorious achievement or service, as opposed to wounds received in action between December 7, 1941, and September 22, 1943, may apply for award of an appropriate decoration in lieu of the Purple Heart."

There are no records of the first individual who received the revived and redesigned Purple Heart. Local posts of the American Legion and the Adjutant Generals of State National Guards both held ceremonies to honor recipients.

What Washington wrote in his Orderly book on August 7, 1782, still stands today: *"The road to glory in a patriot army and a free country is thus open to all. This order is also to have retrospect to the earliest stages of the war, and to be considered as a permanent one."*

Shortly after the award was reinstituted, a group of combat wounded veterans in Ansonia, CT, formed the first chapter of the civilian organization whose membership was composed of recipients of the decoration. Their action gave birth to a fraternal body which, until then, had been but a record on paper. The living organization grew rapidly during and after World War II and is now a nationwide body of men. It became known as the "Military Order of the Purple Heart of the United States of America, Inc." (M.O.P.H.). The organization was chartered by Congress by H.R. 13558, which became Public Law 85-761, on August 26, 1958.

The M.O.P.H. maintains its national headquarters in Springfield, VA, and has chapters throughout the United States. The organization represents veterans' interests before Congress, the Veterans Administration, the Department of Defense, and elsewhere.

In addition, the Order is proud of its key role in the National Service Program. The Order maintains a full time National Service Director who supervises the over 300 salaried and volunteer service officers. All Purple Heart Service Officers have been accredited by the Veterans Administration. They provide assistance and representation for all veterans, their dependents and survivors, in obtaining their rightful entitlements and benefits. All services are FREE.

For additional information on how to join, write or call:

    Military Order of the Purple Heart
    5413-B Backlick Road
    Springfield, VA 22151
    (703) 642-5360

PURPLE HEARTS

Mr. Joseph B. Kelly....GS-11 American Civil Service advisor to the South Vietnamese National Police. Wounded by a mortar attack on the American Embassy in Ben Tre City. While injured he successfully organized the defense of the Embassy. Treated at MACV Hospital. Attack on Embassy classified Secret. No Purple Heart awarded. Second time was ambushed by small arms fire at Tuy Hoa on Hwy 1. He was treated at Tuy Hoa USAF Hospital for a broken back. Even in a back brace he helped the South Vietnamese Nation Police in an counter attack that eliminated 300 Viet Cong saving the lives of many Americans. No Purple Hearts were awarded.

CW-3 Ben R. Games   1st Cav Div. USA aviator was shot down in firelight four times in five months of combat. Awarded the DFC for heroism, Bronze Star, and 13 Air Medals. Lost his spleen, multi ruptures, and was returned to Ft Rucker Hospital. VA changed records to show that he never served in Vietnam. No Purple Heart was awarded.

Copy of Medical Records
Roof over American Embassy quarters of Joseph B. Kelly.

1-9th CAV BLUES DISMOUNT IN A SHAU VALLEY — VIETNAM

6 Nov 1969

'Not A Hero,' Games Says

He learned how to fly in an open-cockpit biplane, flew the super bombers of World War II and jets in Korea, and some 12,000 flying hours later he is flying Chinook helicopters in Vietnam at the age of 45.

At that ripe combat age Ben R. Games, of Union, Mich., is voluntarily flying for the 1st Air Cavalry Division.

Warrant Officer Games, a veteran of the old Army Air Corps and the Air Force, volunteered because "I do not believe any man should have to come back to Vietnam for a second tour unless other men who are trained and capable of doing the job have been over here at least once.'

Games said he hoped that by volunteering he would perhaps delay another pilot's return to the unfriendly skies of Vietnam.

"I know that I'm just a drop in the bucket," he said. "I realize that. I'm no bloomin' hero. But if everyone did it, I think it would make a difference."

He first learned to fly in 1942 in the Army's Steerman biplane. In World War II he flew P-40 fighters in the Far East, then flew P-61's and F-82 night fighters in the South Pacific. Later he was airplane commander for B-17s, B24s B25s, B-26, and B29s.

His pilot's logbook — with its 12,000 hours — reads like a checkerboard history of aviation since before WW II. Besides being qualified in most military aircraft, he has flown nearly every type of private aircraft. In 1965 he singlehandedly flew a single-engine sea plane from Elkhart, across the Arctic ice cap to London, England.

Games is qualified in fixed wing, rotor wing, jet and glider aircraft, and his wife is one of some 100 in the world licensed to fly helicopters.

Now serving with the 228th Assault Support Helicopter Battalion in the 1st Cav, Games spends an average of 10 hours a day at the controls of a CH-47 "Chinook" helicpter, flying supplies to combat fire bases in III Corps.

"In 27 years I've never flown so much," he said. "And I mean fly . . . there's no autopilot, nothing. It's real flying, there's no question about it."

Within 30 days of arriving in country, Games had earned his first Air Medal for combat hours flown. After his 12-month tour in Vietnam he plans to join the Air National Guard at Elkhart.

"I'm not going to quit flying after Vietnam, I just started," he said.

BEN R. GAMES

ARMY TIMES
6 Nov 1969

Helen & Ben Games *** Saigon, Vietnam *** on way to Da Nang

Ben & Helen
in front of our home
Bear Cat 1969

Helen doing
her washing
1969

P. O. Box 1437
St. Petersburg, FL 33731

NOV 3 0 2005
DR. BEN R GAMES
814 CHURCH ST
ELLENTON, FL 34222

In Reply Refer To:
317/CONG/ERR
C 15 227 804
GAMES, Ben R.

Dear Dr. Games:

This is in reply to your letter of November 24, 2005, sent to President George W. Bush. We are writing to you because our office has jurisdiction of your Department of Veterans Affairs (VA) claims folder. We apologize for the delay in responding to your letters.

We will associate your letter with your claims folder. We certainly do not want to discredit your statement of the length of time serving in the military. Our records indicate you have twenty years, ten months of active service and fourteen years, five months of inactive service. Also, our records do not indicate active service from 1962 - 1973. We have requested verification of active service from 1962 – 1973 from the military to verify your statement. If you have discharge papers in that regard, please send VA a copy. We will send you a copy of the service medical records at a later date.

VA only provides the Army with the details of your VA compensation and injuries including a copy of your service medical records. The Army makes the decision regarding CRSC. It is in your best interest to contact the Department of the Army to dispute your issue. If you feel there are additional conditions not presently acknowledged as service connected, or even conditions secondary to a service connected disability, we encourage you to file a claim with the assistance of your designated Veterans Service Organization, The Disabled American Veterans. Any additional disabilities as a result of a new claim are reportable to the Army.

We are sorry that we cannot provide you with a more favorable response to your inquiry at this time.

If you have additional questions or concerns, please feel free to visit our office located at 9500 Bay Pines Blvd., Bay Pines, FL 33744 or telephone toll-free 1-800-827-1000. A Veterans Service Representative (VSR) will be available to assist you. *If you call, please have this letter with you.* Again, you may contact your Veterans Service Organization, The Disabled American Veterans at 727-319-7444 for assistance.

Sincerely yours,

TERRY BERUBE
Acting Director

By Direction of the
Under Secretary for Benefits

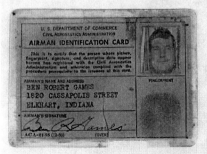
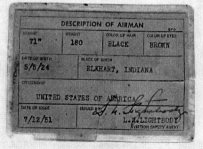

DD Form 214 — Armed Forces of the United States Report of Transfer or Discharge

PERSONAL DATA

- 1. LAST NAME-FIRST NAME-MIDDLE NAME: GAMES BEN ROBERT
- 2. SERVICE NUMBER: W22 189 95
- 3. SOCIAL SECURITY NUMBER: (blank)
- 4. DEPARTMENT, COMPONENT AND BRANCH OR CLASS: ARMY NGUS AV
- 5a. GRADE, RATE OR RANK: CW3
- 5b. PAY GRADE: W-3
- 6. DATE OF RANK: 17 Jun 69
- 7. U.S. CITIZEN: YES
- 8. PLACE OF BIRTH: Elkhart, Indiana
- 9. DATE OF BIRTH: 5 May 24

SELECTIVE SERVICE DATA

- 10a. SELECTIVE SERVICE NUMBER: N/A
- 10b. SELECTIVE SERVICE LOCAL BOARD NUMBER: N/A
- 10c. DATE INDUCTED: N/A

TRANSFER OR DISCHARGE DATA

- 11a. TYPE OF TRANSFER OR DISCHARGE: Released to ARNGUS
- 11b. STATION OR INSTALLATION AT WHICH EFFECTED: Fort Rucker, Alabama 36360
- 11c. REASON AND AUTHORITY: Sec XX AR 635-100 SPN 619 Voluntary Request
- 11d. EFFECTIVE DATE: 2 May 70
- 12. LAST DUTY ASSIGNMENT AND MAJOR COMMAND: H&S Co, Ft Rucker, Ala 3636 0 TUSA
- 13a. CHARACTER OF SERVICE: HONORABLE
- 13b. TYPE OF CERTIFICATE ISSUED: None
- 14. DISTRICT, AREA COMMAND OR CORPS TO WHICH RESERVIST TRANSFERRED: Revert to ARNGUS of Michigan
- 15. REENLISTMENT CODE: N/A
- 16. TERMINAL DATE OF RESERVE/UMT&S OBLIGATION: None
- 17. CURRENT ACTIVE SERVICE OTHER THAN BY INDUCTION
  - a. SOURCE OF ENTRY: OTHER — Ord from ARNGUS
  - b. TERM OF SERVICE: N/A
  - c. DATE OF ENTRY: 1 Mar 69
- 18. PRIOR REGULAR ENLISTMENTS: N/A
- 19. GRADE, RATE OR RANK AT TIME OF ENTRY INTO CURRENT ACTIVE SVC: CW2
- 20. PLACE OF ENTRY INTO CURRENT ACTIVE SERVICE: Union, Michigan
- 21. HOME OF RECORD AT TIME OF ENTRY INTO ACTIVE SERVICE: Route #1, Union, Cass, Michigan 49130

22. STATEMENT OF SERVICE

	YEARS	MONTHS	DAYS
(1) NET SERVICE THIS PERIOD	1	2	2
(2) OTHER SERVICE	26	6	16
(3) TOTAL	27	8	18
TOTAL ACTIVE SERVICE	12	1	12
FOREIGN AND/OR SEA SERVICE	0	7	25

- 23a. SPECIALTY NUMBER & TITLE: 100CU RW Plt Med Trans
- 23b. RELATED CIVILIAN OCCUPATION AND D.O.T. NUMBER: N/A

24. DECORATIONS, MEDALS, BADGES, COMMENDATIONS, CITATIONS AND CAMPAIGN RIBBONS AWARDED OR AUTHORIZED

AM w/13 OLC   DFC

25. EDUCATION AND TRAINING COMPLETED

USAAVNS	HIFC	8 Weeks 1969	Gen Conv
USAAVNS	CH-47	6 Weeks 1969	
USAAVNS	CH-34 Gnd Sch Tng	1 Week 1970	

VA AND EMP. SERVICE DATA

- 26a. NON-PAY PERIODS TIME LOST: None
- 26b. DAYS ACCRUED LEAVE PAID: None
- 27a. INSURANCE IN FORCE: YES
- 27b. AMOUNT OF ALLOTMENT: N/A
- 27c. MONTH ALLOTMENT DISCONTINUED: N/A
- 28. VA CLAIM NUMBER: N/A
- 29. SERVICEMEN'S GROUP LIFE INSURANCE COVERAGE: $10,000

30. REMARKS

Blood Gp: A-
4 Years High School
5 Days Excess Leave: 20-25 Feb 1970
Inclusive dates of service in Vietnam during current period of service:
11 Jan 69 - 5 Feb 70

- 31. PERMANENT ADDRESS FOR MAILING PURPOSES AFTER TRANSFER OR DISCHARGE: Route #1, Union, Cass, Michigan 49130
- 33. TYPED NAME, GRADE AND TITLE OF AUTHORIZING OFFICER: EDWARD W. PARKER, CPT, FA, Actg Asst AG

DD FORM 214, 1 JUL 65

Page 6 — The Colac Herald, Wednesday, February 22, 1967.

Visitors From America

● American visitors to Colac Mr. Ben R. Games (centre) and his wife Helen arrive in Colac after visiting their son in Vietnam.

The couple have been the guests of Mr. and Mrs. W. Riley of Moore Street for several days. Mr. Riley (left) brought them from Melbourne to Colac in his Lake LA-4 aircraft. Mr. Games flew this plane across the Atlantic for Mr. Riley after Mr. Riley had purchased it in America.

● See story of couple's visit to Vietnam and experience in flying on Page 3.

INDIVIDUAL FLIGHT RECORD AND FLIGHT CERTIFICATE - ARMY (PART I)

Period Covered: 1969: Jul - 1970: Jan **Sheet Number:** 107

Name: DAVIS, Don L.
Grade, Br, Component and DOB: CW3, AV, NG, 5 May 24
Type Instrument Certificate(s): STD FW/RW 5 May 70
Originating Organization and Station: Co B, 228th Avn Bn (ASH), 1st Cav Div (Airmobile), APO San Francisco 96490
Operations Officer: WAYNE L. GRIMER, CPT
ARAV: 30 Aug 67

SECTION I - SUMMARY OF PILOT EXPERIENCE

DUTY	FIXED WING			ROTARY WING				OTHER	TOTAL
	SINGLE ENGINE	MULTI-ENGINE		SINGLE ENGINE		MULTI-ENGINE			
		RECIPROCATING	TURBO-PROP	SINGLE ROTOR	TANDEM ROTOR	SINGLE ROTOR	TANDEM ROTOR		
11. AIRCRAFT COMMANDER							372		372
12. INSTRUCTOR PILOT	520	447					34	40	1041
13. FIRST PILOT	908	807		341		345	689		3090
14. COPILOT	3	343		56			38	5	445
15. MILITARY STUDENT PILOT	222								222
16.									
17. CIVILIAN PILOT	3910	2840		240					6990
18. TOTAL PILOT TIME	5563	4437		637			789	734	12160
19. PILOT COMBAT TIME (included in above)								737	737
20. OTHER									

SECTION II - SYNTHETIC INSTRUMENT TRAINER

MONTHS	TYPE	TIME
21. TOTAL THIS SHEET		0
22. TOTAL BROUGHT FORWARD FROM SHEET NO. 106		183
23. TOTAL TO DATE		183

SECTION III - AIRCRAFT QUALIFICATIONS

AIRCRAFT TYPE MODEL SERIES	DATE
O1A	7 Oct 67
UH-1A,B,D	31 Oct 68
OH-13G	11 Dec 67
OH-23B,G	17 Jan 68
U6A	28 Dec 67
UH-19D	10 Apr 68
CH-47A,B	13 May 69

SECTION IV - FLIGHT HOURS ACCRUED - TOTAL HOURS FLOWN BY MONTH (24 Hour Maximum)

JULY	AUGUST	SEPTEMBER	OCTOBER	NOVEMBER	DECEMBER	JANUARY	FEBRUARY	MARCH	APRIL	MAY	JUNE
24	24	24	24	24	24	9.0					

24. REMARKS

Authority to record combat time IAW AR 95-4 and DA Msg 714664, dtd 7 May 65.
Time flown toward accomplishment of annual (semi-annual) minimums:

TOTAL TIME	INST TIME	NIGHT	NIGHT XC	TOTAL XC
494	22	31	6.5	21

Aviator appointed Instrument IP in CH-47A helicopter, Eff date 15 Aug 69, DO 32, Dtd 14 Aug 69.
Records closed this station, 21 Jan 70. Officer PCS to E Bat, 82nd Arty, 1st Cav Div (AM), APO 96490. Authority: VOCO

I HAVE REVIEWED THE ENTRIES ON THIS FORM
GRADE: CW-3   **DATE:** 10 Jan 1970

DA FORM 759 (PART I) 1 MAR 58
DA FORMS 759 AND 759-1, 1 MAR 58, REPLACE DA FORM 759, 1 JUN 56, WHICH IS OBSOLETE.

Ben & Helen
moving day

Bear Cat, Vietnam
1970

family wheels
Honda 50

our new home
Phouc Vinh

INDIVIDUAL FLIGHT RECORD
PILOT

WD AAF FORM NO. 5 (Rev. 1 Oct. 45)

(1) SHEET NO. 26
(2) PERIOD NO. Oct YEAR 1949

PREPARING ORGANIZATION
(3) AF OR COMMAND: 20th AF
(4) BU (GROUP OR SQ): 4th Ftr Sq AW
(5) STATION: APO 239 Unit 2

(6) ORIG. RATING & DATE: Pilot 5 Dec 43
(7) PRESENT RATING & DATE: Same
(8) TYPE INST. CERT. & DATE: White 1 Oct 48

(9) NAME: LAST: GATES FIRST: Ben MIDDLE: R
(10) YEAR OF BIRTH: 5 May 24
(11) GRADE: 1st Lt (R=AD)
(12) A.S.N.: AO 702531

TABLE I — CLASSIFICATION OF FIRST PILOT FLYING TIME

DATE	AIRCRAFT TYPE MODEL SERIES	MISSION SYMBOL	NO. LANDINGS	FIRST PILOT P	INSTRUCTOR IP	DAY CONTACT	DAY WEATHER INSTR. W	DAY OVER THE TOP OT	NIGHT CONTACT N	NIGHT WEATHER INSTR. NW	NIGHT OVER THE TOP NOT	HOOD H	COMMAND PILOT C	CO-PILOT CP	TO	FROM
13	14	15	16	17	18	19	20	21	22	23	24	25	26	27	28	29
4	F61B	S16	1	2:00					1:00	1:00						
6	F61B	T35	1	2:10		:10					2:00			3 OCAs		
7	F6CA	T26A	1	1:10		1:10										
10	F6CA	T26A	1	:55		:55										
21	F61B	S16	1	2:20					1:20	1:00						
				CLOSED OUT:												

(30) TOTALS THIS SHEET			9	3				2	2	2					
(31) BROUGHT FORWARD FROM SHEET NO. 25			1230	663	1434	60	16	165	16	6	198		189		
(32) TOTALS TO DATE			1239	663	1434	60	16	167	16	8	200		189		

(33) TYPED NAME OF OPERATIONS OFFICER CERTIFYING
(34) CERTIFIED CORRECT - SIGNATURE OF OPERATIONS OFFICER
(35) GRADE

INDIVIDUAL FLIGHT RECORD
PILOT

AAF FORM NO. 5 (Rev. 1 Oct 45)

(1) SHEET NO. 29
(2) PERIOD NO. Jan   YEAR 1950

PREPARING ORGANIZATION		(6) ORIG. RATING & DATE Pilot 5 Dec 43	(9) NAME: LAST Games FIRST Ben MIDDLE R.
(3) AF OR COMMAND 20th AF	(4) BU (GROUP OR SQ) 4th Ftr Sq AW	(7) PRESENT RATING & DATE Same	
(5) STATION APO 239 UNIT 2		(8) TYPE LAST. CERT. & DATE White 1 Oct 48	(10) YEAR OF BIRTH 5 May 1924 (11) GRADE 1st Lt.(READ) (12) A.S.N. AO 702531

TABLE I

DATE	AIRCRAFT TYPE MODEL SERIES	MISSION SYMBOL	NO. LANDINGS	CLASSIFICATION OF FIRST PILOT FLYING TIME									COMMAND PILOT C	CO-PILOT CP	TO	FROM
				FIRST PILOT P	INSTRUCTOR IP	CONTACT	DAY WEATHER INSTR. W	OVER THE TOP OT	CONTACT N	NIGHT WEATHER INSTR. NW	OVER THE TOP NOT	HOOD H				
14	15	15	16	17	18	19	20	21	22	23	24	25	26	27	28	29
4	F020	I-26	1	0:45		0:45										
5	T61B	B-10	1	2:00	2:05	2:05	2:00								TACHIKA	NAHA
9	F020	B-10	2	2:40		2:40									ITAZUKE	TACHIKA
10	F020	B-10	1	2:10		0:30		1:40							NAHA	ITAZUKE
11	F51B	B-22	1		2:05	2:05										
17	F020	I-12	1		1:00	1:00										
20	F020	B-16	1	1:55		:25	1:30									
21	F020	B-1408	1	2:15						2:15						
22	F021	B-11	1	1:35		1:10							:25			
27	F020	B-1408	1	1:15		1:15										
28	F020	B-14	1	2:00		2:00										
				CLOSED OUT.												

(29) TOTALS THIS SHEET	17	5	14	4	2	2						
(30) BROUGHT FORWARD FROM SHEET NO. 28	127	668	14	62	18	171	21	8	208		139	
(31) TOTALS TO DATE	129	673	17	66	20	173	21	8	209		139	

(33) TYPED NAME OF OPERATIONS OFFICER CERTIFYING   (34) CERTIFIED CORRECT - SIGNATURE OF OPERATIONS OFFICER   (35) GRADE

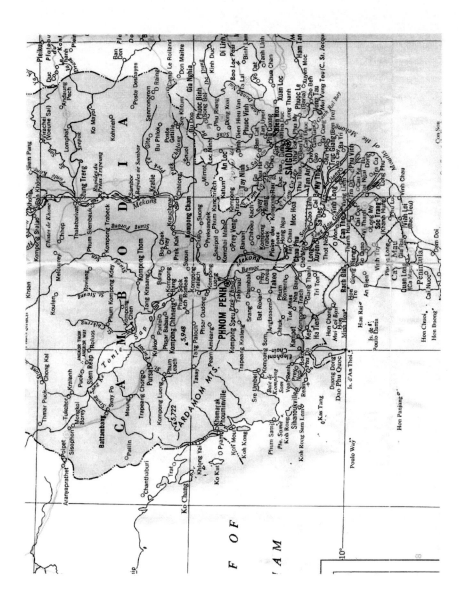

INTERROGATION

VIETNAMESE FABLES AND STORIES

Three generations have passed since the Vietnamese families who fought the Dark Force started making their homes in America. They have worked hard, prospered, and now have grandchildren but have never forgotten their beautiful homeland that was stolen from them. The Vietnamese also brought to America their traditional organization of families and its values. This is a glue that the Dark Force has not been able to overcome and is helping America to remain a shinning light for the world.

During 1969/70 Gentle Ben and Helen lived on a Thai Firebase named Bear Cat near Saigon, Vietnam. While Gentle Ben flew a CH-47 helicopter supporting the Vietnamese PRU Tiger Scouts the Games family became friends with the South Vietnamese government's Minister of Information, Mr. Cat Lee and his family. Today Mr. Cat

Lee's children and grandchildren are successful citizens of the United States, and this nonfiction adventure story of CIA Gilbert H. Moriggia is also part of their history.

No one knows betters than the Vietnamese Citizens of the United States the need to stop Sapper (terrorist) attacks before the killing of children and destruction of family homes becomes a fact of life. The Vietnamese people have learned from their fathers and mothers what it is like to live in a land where a mortar attack at night is as common as rain.

During 1969/70 Helen and Gentle Ben lived on a Thai Firebase named Bear Cat, Vietnam. They heard many Vietnamese stories and legends. This is one of their favorites fables. Vietnamese author unknown:

THE BLIND SON-IN-LAW

There was once a handsome young man who had been blind from birth, but because his eyes looked quite normal, very few people were aware of his affliction.

One day he went to the home of a young lady to ask her parents for her hand in marriage. The men of the household were on their way to the rice fields, and in order to demonstrate his industry, he asked to join them.

He was able to do his share of the day's work but when the men hurried homeward for the evening meal, he lost contact.

When the guest did not appear, the future mother-in-law said, "Oh, that fellow will be a fine son-in-law for he puts in a full day's labor; but it is really time for him to stop for the day. Boys run out to the field and tell him to return for supper.

The blind man heard their voices as they passed by. He followed them back to the house. At the meal, the blind man was seated next to his future mother-in-law, who loaded his plate with food, but then disaster struck. A bold dog approached, and began to eat the food from his plate.

"Why don't you give that dog a good slap?" asked his future mother-in-law, still oblivious to the man's affliction. "Why do you let him eat your food?" "Madam," replied the blind man, "I have too much respect for the master and mistress of this house to dare strike their dog."

"No matter," replied the worthy lady. "Here is a stick. If that dog dares bother you again, give him a good blow on the head."

Now the mother-in-law to be saw that the young man was so modest and shy that he seemed afraid to eat, and would only take small items of food from his plate. To encourage him she selected some sweetmeats from a large platter, and placed them before him.

On hearing the clatter of the chopsticks against the plate, the blind man thought the dog had returned to annoy him. He took up the stick, and gave the poor woman such a blow on the head that she fell unconscious.

****

At this point in the story all the Vietnamese PRU Tiger Scouts would be laughing so loud and hard that Gentle Ben could never understood the end of the fable. In a Vietnamese household the mother or woman is the head of the house, so maybe this is what made the story so funny. Helen rated it a 5, Gentle Ben rate it an 8, and maybe it would be a 10 for the Tiger Scouts.

****

The Vietnamese war was part civil war and part war of the worlds. The real battle was between the Dark Force and the Bright Force. It became a training ground upon how to defeat the United States without fighting on the battle field. It became a part of a political struggle for power in the United States and the world. History records that the Roman Army was defeated by the hoards of barbarians that swept through Italy.

The Vietnam war would be the last war where an enemy would identify itself by declaring war and announcing here I am. From this day on the enemy would hid amongst the citizen of the Bright Force. No man, woman, or child would be safe. The goal of the enemy would be to destroy the morality of the American people. The enemy could live in the United States keeping their women in slavery and teaching that under the freedom of religion that multi slaves or wives was reserved to those who followed the Dark Force.

Understanding the Vietnamese families was critical to devising ways to protect them from the enemy. At this period in America's history mass weapons of destruction was not a deterrent. A military trained to fight a known enemy or other Army would not work. In Vietnam it was the Central Intelligent Agency (CIA) who figured it out and used the enemy against itself. Without these CIA Patriots the United States would have suffered the greatest military defeat in America's history.

SHANGRI-LA VIETNAM

by Ben R. Games, PhD

A long time ago in a land far, far away young men flitted around trees and villages. They flew over rice paddies and amongst rubber plantations. In the jungle they were like humming birds feeding on flowers. Shangri-la you are here.

The sun was bright and the air warm. Flowers brightened the dark green grass, and the land was peaceful. Hats of woven grass protected the men and women working in rice paddies while the young men watched from the sky. In the villages men in white silk shirts and dark trousers walked with women wearing white silk pajama pants under split silk dresses with the bright colors of a rainbow. The picture brought thoughts of love and home to the young men flying their screaming birds. Shangri-la you are here.

Young men of yesterday flitted around the towns and fields with the sound of the blades going "Wop, Wop". Voices in their ears from LZs were calling, "Come to me! Come to me!" It was like flowers calling to Bees, "Pick me! Pick me!" Shangri-la you are here.

Fly like a bird close to mother earth, peek around a building, look over a wall, stop and back up or turn around. Never having to pay for fuel. Shangri-la you are here.

Suddenly a dark wall appears creeping steadily across the land. The fields are empty. The noise of children and bright clothing are gone from the towns. A flash of light, the darkness, the sound of thunder or is it the guns. It's all the same as the young men fly to the battle as moths go to a flame. Many are consumed. Wherever the young men touch the wall of hate and fear, hope springs forth. Shangri-la where have you gone?

FUUJIN

by Ben R. Games, PhD

The Great One created the Heaven and Earth. Then he built a Garden of Eden for his likeness. One day he rested. The next day he looked down upon his garden and saw that the trees, flowers, and meadows needed water for they were turning brown; so he created soft billowy clouds to rain upon them. When he looked down again he saw that the clouds had never moved or the rain stop. The ground had too much water and was turning into mud. God called the Archangel Michael, and told him to think of some way to help His garden grow.

The Archangel formed a committee of Angels and told them that the Lord had said to help the garden grow. They decided that someone would have to move the clouds to where the rain was needed, help pollinate the flowers, and help spread the seeds of all the plants. When the Lord heard of the plan he called out; "FUUJIN".

On warm days I gently rock the baby's cradle, dry the sweat from a man's brow, and move the clouds where they are needed to water the crops. I help move ships, and make high waves for the surfers. I help powered parachutes and trikes fly. I even help sail boats to win races. I make dust devils in dry fields and play with leaves in the fall. I am "FUUJIN'"

Upon God's order I parted the waters of the Red Sea. I can change the shape of Mountains, and create floods. I am on duty all the time. I seldom rest and I can change the winner of sail boat races just for fun. I can destroy crops, towns, or just one building. I can make a fisherman laugh or cry. I am "FUUJIN".

Some call me Typhoon, Hurricane, Tornado, Chinook, Thor, but no matter what I am called everyone knows me, loves me, hates me, and fears me. I am in charge of making the clouds move, and in helping God's children take care of the earth. I am the Okinawa God of Wind, I am, "FUUJIN".

THE CITIZEN SOLDIER

by Joseph B. Kelly

God and soldier we adore
In time of need
And not before.
The war ends and all things are righted
God is forgotten and the soldier slighted.

Not in words are battles won
But by the blood of our countrymen.
Neither will rightist be done
When men of wealth still have their fun.
For when blood is shed on battle field
Rich men get richer from its yield.

For it is he who is well-to-do
Who stays at home
Who's battle you do.
When at last peace comes again
He is the last to give a helping hand.
Yet he is first to pick a fight
And calls you to make it right.

FREEDOM

by Ben R. Games, PhD

During WWII, Korean Police Action, Berlin Crises, and the Vietnam War many of the historical records that included combat injures, photographs, and documents were censored, destroyed, lost, changed, or classified secret to prevent the truth from being recorded in the book of life. Even today some Politicians including Dictators, Kings, Presidents, and Geo-companies (multinational companies operating across national boundaries) are trying to rewrite history. It is an attempted to gain power, to prevent the loss of turf, money, or to enforce a religious belief. Freedom is not just a word.

Whatever the reason those individuals must have forgotten that today is tomorrow's history, and what has been done can not be changed. What they do today is the only thing that will change history. God has given everyone a choice, and this right must be defended. Freedom is not just a word.

During the Vietnamese War period 1962 to 1973 there were two powerful forces fighting to control the minds of the American people. The Dark Force supported by Communist around the world, and the Bright Forces defended by the citizen soldiers of the United States. Vietnam was the battleground but its people were only pawns in the war. Freedom is not just a word.

Some leaders of the most powerful military force on the world tried to conduct their actions during the war with covert operations. The CIA developed its secret Dai Phong Program, the Navy's secret US Operations Mission (USOM) Bangkok, Thailand, the US Army's Special Forces secret support of the PRU Tiger Scouts, and the USAF secret operation at Phou Pha Thi, Laos. All the operations ended in disaster except for the CIA Dai Phong Program, and it became an embarrassment to the American State Department as every city where it was implemented the Communist surrendered, were destroyed, or became PRU Tiger Scouts fighting Communism. The Dark Force flooded the world with lies, misinformation, and sappers (terrorists). America gave up. Freedom is not just a word.

The Vietnam War became a Communist training ground for the world on how to defeat the United States and never have to win a battle. Learn how to replace the Anti-Christ Peace sign with a dollar sign. Use the wire services, internet, and TV to misinform, promote lies, and support politicians who do not believe that America is the land of the Bright Force. Use Cyber-terrorism to turn the people away from honoring their citizen soldiers. Support the idea of developing large super sized bases as a means to save money. Convince Americans that there is no need to train or give modern weapons to citizen soldiers (National Guard/Reserves) as super bases can protect them. Later the military bases can be destroyed with atomic bombs, and the people will give up. It will save money. Freedom is not just a word.

Congress produced a Privacy Act that became law and made personal medical records private information. There are penalties for anyone obtaining and using this information without a release from the individual. It is a good law but the Veterans Administration who has access to all veterans records realized that it was a way to delay benefits and save money. Like an insurance company they could review the data and determine the number of veterans who would die per day from serving America. It is a beautiful diabolic plan. Limit the time a doctor can spend on each diagnosis and force them to use guide lines that will give results projected by computer analysis. Treat the medical symptoms but avoid making a VA diagnosis. No diagnosis no benefits. Freedom is not just a word.

When a United States Senator or Representative needs a question answered there should be no delay, but it takes great skill to write a letter with no-answer to the query. The Veterans Administration has made the representatives of the people into a message center by providing an office in the Rayburn House Office Building B-328 where questions can be asked and delays developed. The VA also employs professional writers or service officers who can twist the truth, change a record, or mislead congress. There are Appeal Boards and legal methods established but the VA is responsible to obtain the documents and veterans medical records for the Appeals Board. To help delay hearings for years the policy is to request other government agencies to verify records. Because these records maybe classified secret the agency holding the records can not verify or provide the information. When the VA is forced to provide the true facts they talk about their budget. The Department of the Army Review Boards Agency wrote the author stating that the Veterans Administration operates under their own laws, rules, and regulations and act as if they are immune to the will of the people. Freedom is not just a word.

Terrorist have discovered that the VA Computers can be compromised and it takes little effort to change a soldier's file just enough to deny claims especially as delay is already the policy. The USAF experts have stated that Cyber-terrorism is now upon us, and it's a war against all citizens of the United States. There is truly a Dark Force and a Bright Force fighting to control the minds of all Americans. Freedom is not just a word.

The VA has made one big mistake for most of their doctors will not accept some pencil pusher or bean counter telling them how to treat an injured soldier. Congress is also catching on and is even now attempting to write laws to regain control. One way would be for congress to reinstate the WW-II draft law and change it to include women, Make the law apply only to VA administrators. Classify those with ten children or more and over the age of 66 as 4F. Draft the others into the Army and send them to the furthest combat theater for a period of the duration plus six months. If they are not killed or maimed when they return home some may finally understand that no American soldier stands alone. Freedom is not just a word.

The Veterans Administration is needed. It is the organization best suited to help the citizen soldiers who were not afraid to fight to save America's Freedom. It is an organization that is supported by American and paid for by Americans. It is not welfare but a promise that Americans have made to each other. Freedom is not just a word.

America's Freedom is threatened by an Unseen Unconventional Arms Race. The Department of Defense, with one of the largest computer networks in the world, has been facing an arms race with an elusive adversary---computer hackers, according to Army Colonel Carl W. Hunt.

Ina press briefing on cyber-security in Washington D.C in November 2005, three panelists discussed the complexities of "cyber-terrorism," a new form of warfare that in recent years has threatened the security of the Defense Department cyber-world.

The USAF, in June 2005, was targeted in a cyber attack that apparently exposed the identities of 33,000 airmen, mostly officers. No identity theft has been reported, although the incident remains under investigation.

Mr. Tom Kellermann, chief knowledge officer of Cybrinth, an e-security and data consulting firm, asserted that 56 million Americans have lost their identities to hackers. America's heavy reliance on technology has left them vulnerable to foreign hackers, who represent 80 percent of attacks, according to Kellermann.

Both Hunt and Kellermann stressed the importance of developing self-healing or self-maintaining computers to monitor all foreign and domestic activities, a technology currently being researched by Microsoft, Sun Systems, and Hewiett Packard, among others. Freedom is not just a word.

35 YEARS AFTER GILBERT

Joseph B. Kelly, T/Sgt USAF and Sarah are now retired. After Gilbert, they served a Christian teaching mission in Argentina for the LDS Church, and later as missionaries in Guatemala microfilming the National Archives.

Joe enlisted in the USAF at the age of 17 in December 1947. After Basic training he became a parachute rigger with 65 jumps, next he became a Survival Specialist, then an Escape & Evasion Specialist before working with Full Pressure Suits for B-57E pilots. The USAF also had him teach scuba diving and snow skiing for survival of US Air Crews. During this same period, he developed 13 inventions or modification of AF equipment to support the USAF fighter aircraft missions.

Joseph B. Kelly was awarded two Air Force Commendation Medals, and three Presidential Citations. His Squadron won the World Wide Weapons Championship, and he worked in special operations for high altitude aircraft of the 7407th Support Group (U-2 & B-57F) in Germany.

After 20 years, he retired from the USAF and accepted an assignment as Gilbert H. Moriggia, CIA Pseudonym, in Vietnam. As Gilbert, he was in charge of liaison and guidance of Vietnamese special police organizations and supervised PRU, SCG, & RDC programs. Gilbert managed these programs, with direct supervision of large complements of indigenous and American subordinates. This came with complete accountability and logistical support for the assigned missions.

The stories and actions of the following individuals during the Vietnam War were interwoven with the story of Gilbert H. Moriggia and became part of the biography of Joseph B. Kelly:

Navy LT/Commander John McCain is now a United States Senator (R).

CW-4 Ben R. Games, PhD, US Army retired is now an author.

S/Sgt Geoffrey T. Barker, US Army, Infantry is now a Lt/Colonel (Ret).

Sgt David L. Walker, US Army Infantry, is now a building contractor.

The war against terrorist is still being fought today only now the battleground is worldwide. During the Vietnamese War, the enemy learned how to use suicide bombers as a primary weapon. They also learned how to flood the media with lies to mislead and destroy the moral of the American people. Cyber-terrorism has become a new form of warfare that is confusing the Media causing them to argue about peace, and the joy of smoking pot.

In Vietnam, it was the body count used by McNamara to measure the success of a battle. Today it is the dollar sign that is used to indicate success, and both are tools of the Dark Force. When Chinese astronauts walk on the moon, will Americans shrug saying that it just costs too much to be free?

To verify Gilbert's story the author researched many documents, journals, attended talks from men who had been prisoners at the Hanoi Hilton, read published letters of Senator John Kerry (D), read Senator John McCain (R) biography, read the written quotes of Hanoi Jane Fonda, met with David L, Walker, Lt/Colonel Geoffrey T. Barker, Joseph B. Kelly, and interviewed Vietnamese families who escaped when the Americans gave up the fight. After thirty-five years he is convinced that;

> That there is a Dark Force and a Bright Force. Also that the war against the terrorist is actually a battle for the Souls of the American People.

CREDITS

Credit for this biographical adventure story "Confession of a CIA Interrogator" must be given to the all those who made it possible.

> All the Military Veterans who fought in the Vietnam War. Citizens Soldiers, National Guard, Draftees, Reserves, CIA Jewelers, CIA Contract Agents, Red Cross, Missionaries, ARVIN, PRU Tiger Scouts, and the Montagnard Sixth Tribe, They are all heroes and part of America's history.

A special thanks to Harry Alexandra, OSS, who taught the author about the need for gathering foreign intelligence, and to the CIA patriots who help protect America today. History is a window that looks into the future.

Ensign Marie Tillery, Staff Assistant Veteran Affairs for Senator Martinez, WO1 Mistie Kay Wisniewski US Army (CRSC), and many others helped in the writing of this story. Each in their own way are still helping to protect American families from the Dark Force.

To these men and women; Thank You For Helping America!

Boston Whaler used by Gilbert with M-60 machinegun & two 40 HP motors

AUTHORS PERSONAL INFORMATION

www.GamesLittleBigBooks.com

Ben R. Games, PhD, Major, CW-4, TCNA-6, flew bombers and night fighters during WWII. Then Jet Fighters for the USAF during the Korean War, and Chinook helicopters in Vietnam for the 1st Cavalry Division. He is a member of the North American Mach Busters Club and of the Distinguished Flying Cross Society with 737 recorded combat hours. After 35 years he retired from military flying and became the manager of the Turks & Caicos National Airline.

He served in Vietnam as a pilot with the 228 Aviation Battalion, Company B, 1st Cavalry Division, and is a life member of Army Aviation Class 43K, 1st Cavalry Division Association, MOAA, USAF Association, VHPA, DFC Society, National Guard Association of the US, Camp Graying Officers Club, VFW, American Legion, and the DAV.

During his military service Ben was awarded the Distinguished Flying Cross, Bronze Star, 13 Air Medals, Army Commendation Medal, 2 Medals for Valor, and two Legion of Merit.

During the past fifty years stories of his flying adventures have been read by people around the world. The stories are semi-autobiography in nature or historical adventures.

AUTHOR OF: SOLO 90-Z, My Guardian Angel, Without Prejudice, BEYOND, Little Big Books sub titles, Powered Parachute Zone, Montana's Cruising Vacation, Poems & Stuff, Sinking of the Carol "B", Santa's Secret, Death of a Patriot, Adventures of Benny Bob, and Confession of a CIA Interrogator.